T0224565

Electronic Visual Music

'This offers an original take on visual music, rooted in the author's extensive experience working as a creator in the field. It has a distinctive focus on electronic music and visuals, and offers a useful mix of historical antecedents, interviews with current practitioners and pointers for practical and creative work.'

Joseph Hyde, *Emeritus Professor, Creative Music Technology, Bath Spa University*

'Dave Payling, due to his large experience as a professor, researcher and artist, successfully accomplished the task of producing a comprehensive book on the fascinating theme of electronic visual music. He addresses didactically the relevant theoretical topics and provides fertile insights related to the creative process for those interested in diving into this art.'

Antenor Ferreira, *University of Brasília, Brazil*

Electronic Visual Music is a comprehensive guide to the composition and performance of visual music, and an essential text for those wanting to explore the history, current practice, performance strategies, compositional methodologies and practical techniques for conceiving and creating electronic visual music.

Beginning with historical perspectives to inspire the reader to work creatively and develop their own individual style, visual music theory is then discussed in an accessible form, providing a series of strategies for implementing ideas. Including interviews with current practitioners, *Electronic Visual Music* provides insight into contemporary working methods and gives a snapshot of the state of the art in this ever-evolving creative discipline.

This book is a valuable resource for artists and practitioners, as well as students, educators and researchers working in disciplines such as music composition, music production, video arts, animation and related media arts, who are interested in informing their own work and learning new strategies and techniques for exploration and creative expression of electronic visual music.

Dave Payling is an audio-visual artist based in Staffordshire, UK. His compositions have been performed at events including ICAD: Sydney Opera House, ICMC: Shanghai and Understanding Visual Music: Brasilia. He is a section editor for *Dancecult: Journal of Electronic Dance Music Culture.*

Sound Design
Series Editor: Michael Filimowicz

The *Sound Design* series takes a comprehensive and multidisciplinary view of the field of sound design across linear, interactive and embedded media and design contexts. Today's sound designers might work in film and video, installation and performance, auditory displays and interface design, electroacoustic composition and software applications, and beyond. These forms and practices continuously cross-pollinate and produce an ever-changing array of technologies and techniques for audiences and users, which the series aims to represent and foster.

For more information about this series, please visit: www.routledge.com/Sound-Design/book-series/SDS

Electronic Visual Music

The Elements of Audiovisual Creativity

Dave Payling

Routledge
Taylor & Francis Group

LONDON AND NEW YORK

Designed cover image: © Dave Payling

First published 2024
by Routledge
4 Park Square, Milton Park, Abingdon, Oxon OX14 4RN

and by Routledge
605 Third Avenue, New York, NY 10158

Routledge is an imprint of the Taylor & Francis Group, an informa business

British Library Cataloguing-in-Publication Data
A catalogue record for this book is available from the British Library

ISBN: 978-1-032-32671-9 (hbk)
ISBN: 978-1-032-32663-4 (pbk)
ISBN: 978-1-003-31613-8 (ebk)

DOI: 10.4324/b23058

Typeset in Times New Roman
by Apex CoVantage, LLC

To Julie, Hannah, Ben and all the canine friends that keep us grounded

Contents

Figures

Acknowledgements

A few brief mentions to those who helped in some part in this book seeing the light. Firstly all the artists who kindly agreed to be interviewed for this book, they provided some great insights and made compiling the second chapter an easy process. To Tim Howle for guiding me through my PhD and helping me shape my audio-visual experiments into something that had academic merit. Thanks to Jo Hyde for his endorsement and being the first person to suggest expanding my PhD thesis into a book. To Antenor Ferreira for providing the endorsement and always being a supportive force in Brasilia. Calum Wilton proofread a couple of chapters and provided valuable feedback. Jon Weinel gave me some early advice and recommendations in moving forward with the book idea. To Mark McKenna for friendly discussions on the publishing process and general encouragement. Thanks to all the people who have supported NoiseFloor, there are many I could mention - you know who you are. And always thanks to all the students at all levels and stages of their journeys for constantly questioning and presenting new perspectives.

Introduction

Electronic Visual Music

Introducing Electronic Visual Music

Welcome to my book about visual music. It has been a desire of mine for some time to commit to print my thoughts about this subject and consolidate some of my experiences, both in teaching at university and in my personal creative practice, which extends back to the 1990s. The motivation to write has also been inspired by reading the many excellent books and articles about visual music, some of which are referenced in this book. Visual music and audio-visual media continue to grow as disciplines and there is a wealth of high-quality writing and artworks available; hopefully I can contribute something useful and meaningful from my own perspective. Writing has been both an enjoyable and challenging process. My own thoughts and preconceptions have been challenged every step of the way. Only when committing the words to paper did it become apparent that something lacked the clarity I thought it contained. This has encouraged me to try and reinforce every discussion by undertaking more reading and reflection. I hope this has been mostly achieved, but welcome feedback and further discussion so updates can be made when this book is revised at some point in the future.

The topic is visual music, a creative artform, but writing itself is also a creative process, one that I am still learning. It is necessary to work on the fine detail, the intricacies of each word, sentence and paragraph and their emerging meanings before returning to an overview of each chapter and the whole book to objectively evaluate its progress. The process is not unlike composing a piece of music or visual music, only the medium is different, and the artist will understand that composition rarely goes as originally intended. The enjoyable aspects are seeing things come together after days and weeks of working. Refreshing one's knowledge in a subject that has been studied for so long still reveals new ideas, artists and authors that feed into the discussion and inspire one to create new pieces and articles. Writing is an excellent way to understand a subject.

This book is very much my own perspective on visual music, albeit being informed by the interviews in Chapter 2. It is a subject that has a relatively recent history but one which has become more established and diverse. It is not necessarily helpful to try and pigeonhole a range of works into one category but sometimes there is no way around this without identifying sub-genres and increasingly

DOI: 10.4324/b23058-1

detailed classification systems. For now, I will refer to artworks relevant to this book as 'visual music' and discuss the rationale for this soon. The golden period of this artform resides in the mid to late 20th century and is associated with artists including Viking Eggeling, Mary Ellen Bute, Oskar Fischinger, John Whitney . . . These artists are the influential pioneers of the art and created defining compositions of the genre. Many contemporary artists have been inspired by these pivotal moments and refined and updated the techniques to create new works of outstanding beauty. My own step into visual music stems from experiences I had in popular music, the use of computers and augmenting live music shows with screened visuals in the 1990s. At that point in time, I did not consider defining the media I was producing as 'visual music', and I still do not consider that work to be classed as such, but those experiences have informed my later practice by providing a good foundational insight into how visuals work as an independent medium and cohesively in conjunction with music. These are important considerations when creating visual music.

These formative experiences were gained in my home city of Sheffield, UK. During the 1980s and early 1990s there was a culture of live experimental electronic music which, by some artists, was accompanied by video projections. These events took place in local clubs such as the City Hall Ballroom, the Arches and a range of smaller venues dotted around the city. The gritty visuals and audio often had an industrial and experimental leaning, at least in part inspired by the industrial heritage of the city. At the time, Sheffield really was a creative place with its own distinct aesthetic and just living in the area meant osmotically absorbing these influences by trading ideas with friends and artists. Some acts typified this approach such as Cabaret Voltaire and Clock DVA, the latter who were a strong inspiration for me to begin working with digital video animation using Electronic Arts' Deluxe Paint on the Amiga computer. This seemed to be a natural progression to the MIDI sequencing work I was doing with Microillusions' Music X sequencer. It was the dawn of the home personal computer and multimedia software, and it unlocked many creative possibilities. These formative explorations found their way into the live arena when electronic video was produced to accompany the band I was performing with. The digitally generated visuals were recorded to analogue VHS tape and screened on old and heavy cathode ray tube television sets mounted on top of speaker stacks. Being the car driver meant I was the person mostly responsible for storing and lugging around the equipment, always strenuous work! The videos in these early productions were not intended to be of interest in their own right; they were more a way of enhancing the musical performance in a visual way. Something else for the audience to engage with. The videos were abstract and colourful and often behaved more in the manner of a light show, rather than containing any kind of narrative or representation of the musical content. The audience seemed to accept and enjoy the visuals and responded with positive comments about their inclusion in the performance. Preparing and creating all this additional media was time-consuming, but it provided an additional layer of interest for the band and the audience, which set us apart from many of our peers who, apart from the acts noted above and a few others, were still largely performing music-only stage shows.

Some of the ideas concerning the use of abstract colour and light experienced in these early videos have since been carried forward into my later practice and theories related to composition and performance. Thomas Wilfred's Lumia (1947, 247), for example, seems particularly relevant in this respect and was used as inspiration for *Space Movement Sound* (Payling 2010) and *Diffraction* (Payling 2012), the final two compositions of my PhD portfolio, as well as being a more generalised influence in later works.

A further advancement to later work was the intention for the visuals to become more cohesive with the music. In an effort to achieve this, my artistic output since 2008 has explored the workflow and techniques used in creating what I identify to be closely aligned with the artform of visual music. This genre is focused and refined to enhance the synergy between music and image, to achieve a more unified balance between the two. The technique of 'compositional thinking' (Hyde 2012, 171) was used in part to achieve this. In compositional thinking the compositional form of music influences that of the visual. This is where one connection between sound and image can begin. The primary technique to further enhance these relationships however, was 'material transference' (Hyde 2012, 170). With this technique, the attributes of one medium are transferred by some means to the other. So, for example, visual colours could inform the choice of musical timbres. In my earlier portfolio pieces material transference was achieved through automated techniques and parametric mapping, the mapping of parameters from one medium to another. Sound amplitude modifies shape size, for example. Aside from these rigid mapping processes, a more subtle transference was applied, through the composer's influence and intuitive interpretation between the media. This occurred more frequently in later work as confidence in composition increased, but in all approaches the artist's guiding hand made the final judgement. These are only a few examples of approaches that can be used during composition and performance, many others will be discussed throughout this book.

Intended Audience

Although compiled with newcomers and existing visual music practitioners, audio-visual composers and academics in mind, this book may be of specific interest to those who come from a music background and who are considering adding a visual component to their work. Although I think of myself as an interdisciplinary artist, my experiences originate in popular music composition and performance and later in more experimental musical forms. The visual skills came later and have been developed alongside the core practice of music making. The discussions about visuals in this book are therefore written from this perspective; a musician seeking meaningful use of visuals. This may also be true of others in a similar position. The necessity to diversify roles, the prevalence of video-sharing sites and a desire to have as much artistic control as possible means many musicians have an interest in working with visual materials. It is hoped therefore that this discussion is of use to those readers with similar ideals. This may of course work the other way. Visual artists, who are interested in adding a musical dimension to their creations, should hopefully be able to derive pertinent information and techniques from this work. Music is the prevalent undercurrent which drives much of the discussion, and some practical techniques in sound will be introduced in Chapter 5.

Sound and Image Genres

Visual music has, in recent times, diversified and become a more nuanced practice. Originally identified with the types of film produced by the originators of the artform introduced above, it can now be used to describe a range of different sound- and image-based productions. Some related terms that may be used in a similar context are audio-visual, videomusic, multimedia and intermedia. Intermedia is a term used by Kapuscinski, where he describes works 'in which sounds and images are given equal importance and are developed either simultaneously or in constant awareness of each other' (Kapuscinski 1998, 43). This is a good contender for describing the process of creating visual music, but the label doesn't specifically refer to sound and image. Videomusic is the name adopted by Jean Piché and is very apt for the compositions produced by him and the artists he works with. It is quite closely connected to the Montreal scene so is not considered as fitting for a more generalised discussion. Louise Harris (2021) makes a strong case for using the name 'audiovisual' (omitting the usual dash in audio-visual), arguing that both media are composed together to create cohesion, whereas in visual music the visual conforms to the musicality of the work. I do also occasionally use audio-visual when referring to media in this genre, but this has some distracting associations. First, the term 'audio-visual' has a usage unrelated to the artistic practice itself. Audio-visual can refer to the apparatus, equipment, projectors and loudspeakers involved in sound and vision reproduction. It is a means of screening selected works, not the creative process itself. In isolation the word 'audio' refers to transmitted and recorded sound. Although there are techniques where all sound can be structured as music, there are many uses of the term audio that do not refer directly to music. The use of the word 'music' therefore promotes sound from its functional capabilities to something more associated with art. Although the word 'visual' has several meanings, when coupled with music to form the phrase 'visual music' it forms a harmonious combination reflective of the artworks in this genre and appropriate to the many artists and compositions discussed in this text.

Defining Visual Music

For the purposes of this book, visual music involves the portrayal of musical qualities in sound and image media. One of the first definitions and implementations of this artform related only to the visual medium, originally paintings; this type of visual music did not require sound. In this discussion, however, music is a key component of the whole, hence the references to sound and image. From a visual perspective visual music would normally predominantly include time-based visual imagery with underlying musical form or qualities. Even if this is not the intention at the outset there are several techniques that, once applied, will naturally impart what could be considered as musical qualities. The use of metric montage (Eisenstein 1949, 72–73), for example, will impart a musical time base to moving images. Visual abstraction is another trait of many visual music creations and from a musician's perspective, abstract visuals may have certain traits and behaviours which could be interpreted as reflecting musical qualities. Although abstraction is not essential in visual music, many of the pieces discussed in this book are non-narrative and non-representational in nature. They are also accompanied by sound; therefore complying with Evans' (2005) definition of visual music.

Electronic Visual Music

Visual music is one part of the book title; prefixing it with the word 'electronic' could be a contentious move. The inclusion of 'electronic' stems from my personal interests in electronic music and by extension, electronic visuals. Visual music historically originates from painting and direct abstract animation using canvas and film stock, whereas many recent works, at least those introduced in these pages, are created with the aid of computer technologies. I personally have a very heavy reliance on computer-based techniques to realise my projects and feel the term electronic fits more closely with this approach. This does raise the question, why not 'digital visual music'? To which my reply is that electronic music comprises a recognised range of related music genres, whereas digital music is more frequently associated with production techniques and delivery formats. I therefore propose *Electronic Visual Music* as a name that can be applied specifically to my own work but also to similar productions that use electronic means for the production of sound and/or image. Although some artists will prefer their own specific terminology, for the sake of brevity I will mostly refer to their works as 'visual music' or more specifically 'electronic visual music'.

Even though this book identifies with visual music and electronic visual music, I have no qualms with alternative names being used. There are many parallels and similarities between the previously discussed genres and techniques which will become apparent throughout the discussion ahead. Ultimately, I feel anyone with an artistic slant towards sound-image composition and performance is within the same family of practitioners and hopefully they might find some useful discussion in this book. The label being used is not of primary importance; it is the artwork itself that should be prompting engagement and discourse.

Book Structure

The content and order of the book is conceived around the five elements used in various Eastern schools of philosophy. This is perhaps an unusual way to organise the contents of a visual music book, but it aligns with the interests of some of the past luminaries of the artform and has provided a higher-level structure which has helped me produce and organise the written materials. The elements of spirit, air, fire, water and earth are all different states of energy that can be transformed from one to another. The purest form of energy spirit exists all around, it is the element from which all others are derived. Through motion, heat and condensation it gradually solidifies into solid earth 'energy' which, in the case of this book, signifies the transition from inspiration to realisation through Chapters 1 to 5. Considering the transition between states and beginning with spirit, this fine energy can begin to flow as moving air. As the air begins to move more rapidly, heat is generated, creating fire. This heat creates condensation from the air which can distil into water and finally solidify into matter and earth. These elements therefore possess different qualities but all influence and stem from each other, much in the way performance can influence composition for example. Applying these elements to the structure of this book presents a novel and hopefully intuitive approach to ordering the chapters. First, spirit relates to the wider aspects, historical context and influences of visual music. It is the spirit of the artform which can be channelled in future works. Air is used in the communication of ideas by various composers, which are presented as interviews in Chapter 2. These discussions took the form of verbal communications using vocal movement of air, hence their designation as the air element. Fire represents the energy of performance, the stirring of nervous energy and the physical gestures used to create the art. Water represents the creative state of composition, converting ideas into arrangements and tapping into the composer's emotions and intuition. Last, earth presents practical techniques used in creating the artworks, taking the preceding insights and processes and applying practical methods to create the finished piece.

Even though this idea has helped me to conceive and structure the book, so it reads from start to end as a continual developing discussion, it can be explored freely in any order the reader chooses. Apart from reference to specific methods, theories and artists that follow through this discussion, each chapter has sufficient context to be explored as a standalone article. This is especially true of the interviews section, and if looking to get started with some practical techniques one could go straight to the final chapter.

Chapters Overview

A brief overview of the content of each chapter is provided below.

Chapter 1. Discover: Electronic Visual Music in Context – A wide-ranging historical context that examines links between images and music in science and art and how they have been applied. This chapter introduces the luminaries and pioneers of visual music dating from the early to mid-20th century. It documents early attempts to convey musical concepts in visual form to present-day developments and the use of computer-based animation and sound creation to create similar results. It also examines related subjects such as synaesthesia and sonification and how they have influenced interdisciplinary composition.

Chapter 2. Communicate: Electronic Visual Music Conversations. This chapter presents interviews with current visual music artists giving perspectives on their work and techniques of composition and performance. These have been selected from a broad range of backgrounds and techniques, in an attempt to provide a diverse range of perspectives on the subject.

Chapter 3. Perform: Live Electronic Visual Music. The theory and practise of performing with audio-visual media. The theories of liveness, performance and flow state will be discussed in relation to audio-visual performance. The types and roles of visuals in live performance are considered, followed by techniques of parametric mapping and controller conditioning. A case study of a networked and in-person performance will be discussed.

Chapter 4. Compose: Electronic Visual Music. This chapter opens with the various approaches that can be used when composing with sound and image. It discusses techniques of montage, material transference and compositional thinking and how they can be used to create cohesion between sound and image. Spatiality of visual music is addressed by considering audio-visual landscapes and the role of gesture and texture visual music. A comparison is made between algorithmic generative techniques and composer rendition with the inclusion of an algorithmic case study. Thomas Wilfred's lumia artform is presented as a technique for informing visual music composition.

Chapter 5. Create: Electronic Visual Music. This chapter discusses the various technologies which can be used to create visual music compositions and performances. The benefits of each technology will be introduced along with a discussion of their strengths. Various practical video and sound production techniques are then examined and related back to some of the theoretical discussion in earlier chapters. The audio-visual production workflow is discussed with reference to specific mastering and rendering techniques. A composition case study is included detailing how each stage of a project is undertaken, and the final artefact realised.

Thanks again for taking time to read my book. Now on with the discussion . . .

Note: When mentioned in the text, compositions are highlighted in *italics*, for example, *Aquatic*. This is to identify them as creative works and to avoid confusion between composition titles and actual words and terminology.

Online Video Viewing

Throughout the book I have referenced visual music compositions which are hosted mostly on YouTube and Vimeo. One reason for doing this is to attempt to contribute to the number of references creative works such as these accumulate. When compared to written articles, published in journals and books, compositions receive far fewer citations. These practice-based outputs are, however, equally as important to the artistic research community and should gain the credibility they deserve. Apart from this the links are intended to be followed up and viewed by the reader to gain a full appreciation and contextualisation of the discussions. Please take the time to visit the links, view the media and where appropriate comment on, like and share the videos. It will give the artistic community the recognition it deserves and provide motivation to continue.

References

Eisenstein, Sergei. 1949. *Film Form: Essays in Film Theory*. Translated by Jay Leyda. A Harvest Book 153. New York: Harcourt, Brace & World.

Evans, Brian. 2005. 'Foundations of a Visual Music'. *Computer Music Journal* 29 (4): 11–24.

Harris, Louise. 2021. 'Composing Audiovisually: Perspectives on Audiovisual Practices and Relationships'. Routledge & CRC Press. 21 July 2021. www.routledge.com/Composing-Audiovisually-Perspectives-on-audiovisual-practices-and-relationships/Harris/p/book/9780367346911.

Hyde, Joseph. 2012. 'Musique Concrète Thinking in Visual Music Practice: Audiovisual Silence and Noise, Reduced Listening and Visual Suspension'. *Organised Sound* 17 (02): 170–78.

Kapuscinski, Jaroslaw. 1998. 'Basic Theory of Intermedia. Composing with Sounds and Images'. *Monochord. De Musica Acta, Studia et Commentarii* 19: 43–50.

Payling, Dave. 2010. *Space Movement Sound*. Electronic Visual Music. https://vimeo.com/15660433.

———. 2012. *Diffraction*. Visual Music. https://vimeo.com/40227256.

Wilfred, Thomas. 1947. 'Light and the Artist'. *The Journal of Aesthetics and Art Criticism* 5 (4): 247–55.

1 Discover

Electronic Visual Music in Context

Discovering Visual Music

This chapter represents the spirit of visual music. It touches on important artists, related concepts and various approaches to audio-visual art that can serve as inspiration for future works. It is energy in its purest conceptual form that can be transformed into new materials. While some engage with artforms, such as painting and music, as unique means of expression, others work on the premise that there is an innate connection between all arts and even that they share fundamental aspects on a spiritual level. As this book deals with visual and musical matters, it is connections between these arts which are most important for understanding the wider context of visual music. Visual and musical arts' shared ideals have been discussed and documented at least since the ancient Greek philosophers and continue to the present day. One recurring correspondence is that coloured hues can be equated to musical pitch. Isaac Newton touched on this in *Opticks* (1704) when discussing the nature of his experiments with light. When he split white light into the familiar rainbow spectrum, he made analogies between colours and musical pitch. This was a conceptual mapping with the note D at the boundary between violet and red with subsequent colour boundaries mapped to a Dorian mode musical scale (Hyde 2022, 136–37). Other artists and writers extended the colour/pitch concept to include the 12-tone chromatic musical scale mapped directly to coloured hues. Some of the different correspondences have been documented by Collopy (2009) and it is worthy of note how similar some of the choices are. Rather than the allocations being completely arbitrary there are some distinct similarities, especially in the C, D, E and B notes, possibly due to familiarity and intuitive alignment with the rainbow spectrum.

Another relationship that exists between colour and music is the one between hue and musical timbre. Hue has been described as 'the principal way in which one colour is distinguished from another' (Collopy 2000, 356) in a similar way that timbre has been defined as 'the distinctive tone quality differentiating one vowel or sonant from another' (Humberstone 1991, 1613). Furthermore, a common way of referring to instrumental timbre is as the 'colour' or 'tone' of the instrument and linguistically klangfarbe, or 'sound colour', is the German equivalent of the

DOI: 10.4324/b23058-2

English 'timbre'. One study (Scholes 1950) demonstrated a general agreement between musicians for printing musical scores of different instruments in different colours. For example, a red score would be used for brass instruments and blue for woodwind. A similar suggestion by Albert Lavignac (Jones 1972, 111) resulted in a list of instruments and colour associations. Flutes were assigned to blue of the light sky, oboes to a crude green tint, trumpets to crimson or orange and so on. Lavignac discussed these colour associations, explaining how a composer could use contrasting instruments to create a composite piece of music in a similar way an artist would use colours:

> I would add that the art of orchestration seems to me to have much similarity to the painter's art in the use of colour; the musician's palette is his orchestral list; here he finds all the tones necessary to clothe his thought, his melodic design, his harmonic tissue, to produce lights and shadows, and he mixes them almost as the painter mixes his colours.
>
> (Lavignac 1899, 184)

It is possible to map between colour and sound parameters using these relationships, but artists have also attempted to interpret them in a more intuitive way. Wassily Kandinsky was an influential abstract visual artist, based at the Bauhaus Art School in Germany, who desired to express musical qualities in painting. Music is inherently abstract; it cannot represent real scenes in the same way paintings can; it can only suggest certain qualities. Kandinsky attempted to emulate the purity of this aspect of music in his paintings, which have a free-flowing style that omits representational features and instead utilises abstract visual strokes suggestive of musical gestures (Kandinsky 1939; 1923). Colours were used expressively, and he went some way to explain the links he considered to be present between colours and sounds. The other major contribution of Kandinsky was his discussion on colour and form (Kandinsky 1914, 44–78). He considered these as the two components available to a painter that can be employed to produce the desired effects.

> The value of certain colours are emphasised by certain forms and dulled by others. In any event, sharp colours sound stronger in sharp forms (for example, yellow in a triangle). Those inclined to be deep are intensified by round forms (for example blue in a circle). On the other hand, if a form does not fit the colour, the conjunction should not be considered "Inharmonious" but rather as a new possibility and, therefore, as harmony. As the number of colours or forms is endless, the combinations and effects are, also, infinite. This material is inexhaustible.
>
> (Kandinsky 1914, 46–47)

Even though here he was describing visual effects, Kandinsky's discussions often incorporated musical terms. The Swiss artist Karl Gerstner expanded on

Kandinsky's ideas, focusing even more deeply on the emotive and spiritual signifi-
cance of colours . . .

> In our experience of colors and in our knowledge of the laws underlying
> them we penetrate to the ultimates that hold all life together from the origin,
> evolution, and structure of the cosmos to the moral category in which indi-
> vidual colors produce their ethical /aesthetic effect.
>
> (Gerstner 1981, 48)

Gerstner's artistic œuvre reflects these sentiments with many boldly coloured
works. His writings also explored the relationships he believed existed between
colours and sound in a similar way to Kandinsky.

Colour Organs

Colour and performance was brought together in devices known as colour organs.
Although colour organs could take on many forms, they commonly had a standard piano-type keyboard with additional apparatus that allowed the projection of
coloured light. Although many colour organs resembled musical instruments they
were silent devices producing only light and required musical accompaniment if
sound was part of the performance. Colour organs have a long lineage, with the
first assembled and documented device created by Louis Bertrand Castel (1740).
Castel was a French monk with an interest in music and mathematics. His 'Ocular
Harpsichord' (Franssen 1991) was a mechanical device that opened a curtain when
a key was pressed on the keyboard so light could be shone through coloured glass
squares. Castel's Ocular Harpsichord generated some interest but did not appear
to have any long-term success. In 1893, Alexander Wallace Rimington patented
an invention for 'Method and Means or Apparatus for Producing Colour Effects'
(Rimington 1895) which describes the workings of and written detail and series of
schematics instructing how such a device could be constructed. Subsequent inven-
tors used similar principles with the colour organs built by Bainbridge Bishop with
documented in his book (Bishop 1893). Bishop's instrument comprised coloured
glass filters through which light was projected and reflected onto a ground glass
tablet above the keyboard. Chromatic musical pitches were mapped to the rainbow
colour spectrum spanning across the keyboard.

> I had some trouble in deciding how to space the intervals of color, and what
> colors to use, but finally decided to employ red for C, and divide the prismatic
> spectrum of color into eleven semitones, adding crimson or violet-red for B,
> and a lighter red for the upper C of the octave, and doubling the depth and vol-
> ume of color in each descending octave, the lower or pedalbass notes or colors
> being reflected evenly over the entire ground. The whole effect was to present
> to the eye the movement and harmony of the music, and also its sentiment.
>
> (Bishop 1893, 8)

It was also possible to combine colours, when playing a chord for example, and for the intensity of the lights to be adjusted. Bishop described how he believed these techniques could be used to achieve similar effects as music:

> I soon found that a simple color did not give the sensation of a musical tone, but a color softened by gradations into neutral shades or tinted grays did so; also, that combinations of colors softened by gradations into neutral shades or tinted grays, with the edges of the main colors blending together, or nearly together, rendered the sensation of musical chords very well indeed. The impression or sensation of the lower bass notes I could get only by low-toned or weak colors diffused over the whole field, the higher colors or chords showing smaller on this ground.
>
> (Bishop 1893, 5)

This describes how colours can create a musical 'sensation' and how the subtleties of colour variation are more successful in achieving this. Bishop found limitations in a one-to-one pitch to colour mapping; more subtle and complex relationships should be developed. In many ways the colour artform lumia, developed by Thomas Wilfred, permitted this finer level of detail and nuance.

Thomas Wilfred and Lumia

The clavilux ('a name derived from the Latin, meaning light played by key' (Stein 1971, 4)), invented by Thomas Wilfred, was a device that permitted performances with colours. Compared to colour organs, however, the clavilux had additional capabilities allowing the form and motion of coloured light to be manipulated. The three factors of light, form, colour and motion were used to create a new artform Wilfred named lumia (Wilfred 1947, 247). There were several clavilux instruments, some of which were self-operating and designed for unique performances or installations in galleries and others which had greater degrees of interaction. The visual output of the instruments was a light projection of transforming coloured forms with an ethereal quality. The clavilux was capable of creating complex visual arrangements:

> A typical composition contains one principal motif with one or more subordinate themes. Once chosen, they vary infinitely in shape, color, texture, and intensity. The principles evident in plastic and graphic compositions—unity, harmony, and balance—function kinetically in lumia. When movement is temporarily suspended in a lumia composition, the result is a balanced picture. However, the static picture's ultimate meaning can only be seen in relation to what follows it.
>
> (Stein 1971, 3)

Wilfred described lumia in detail; the three principal lumia factors of form, colour and motion each had four sub-factors and he grouped these into a 'graphic equation'.[1] The combination of the lumia factors, to the left of the equation, produced the artistic potential on the right:

> Place this inert potential in a creative artist's hand, supply him with a physical basis—screen, instrument and keyboard—and when his finished composition is performed, the last link in the chain has been forged and you have the eighth fine art, lumia, the art of light, which will open up a new aesthetic realm, as rich in promise as the seven older ones.
>
> (Wilfred 1947, 254)

Wilfred believed he had laid the foundations for a new art that would continue to be studied and practised but there was limited contemporary work using lumia or the clavilux instruments until a revival took place in the early 21st century.

Animated Visual Music

The previous discussion relates to intersections between music and visual art alongside instruments capable of performing abstract visuals. During the early part of the 20th century animation, in the medium of cinematic film, was progressing through its formative years and it was this medium that propelled audio-visuals into a new era of artistry and development. Pioneering abstract animation artists including Walter Ruttman, whose *Lichtspiel Opus I* (1921) is credited as the first screening of an abstract animated film, and Hans Richter and Viking Eggeling, who were connected with the post first world war Dadaist art movement, created minimal abstract films inspired by music. Richter's *Rhythmus 21* (1925) features black and white rectangular images that fade in and out of the screen and Eggeling's *Symphonie Diagonale* (1924) has become a much celebrated classic composition in silent visual music. Animation continued to develop and the practice gradually became more widespread with several artists creating what Oskar Fischinger described as 'absolute, non-objective film' (Fischinger 1947, 1). Absolute is used here to describe a pure type of film exploring techniques of composition and motion and its non-objective nature referring to its use of non-referential abstract imagery. As physical film roll was the animation medium, it was viable to hand draw, or paint, individual frames directly onto the celluloid or draw animations on other media to later be photographed onto the film. William Moritz titled Fischinger's biography *Optical Poetry* (Moritz 2004), based on one of Fischinger's films, *Optical Poem* (Fischinger 1937). This describes the poetic nature of the animations Fischinger created, which are still shining examples of poetic beauty to the present day. Fischinger also experimented with drawn sound by adding images to the sound strip of the film to synthesise music from pictures, a technique that will be returned to shortly.

Another artist, Len Lye, who originated from New Zealand, created animations using similar direct filmmaking techniques. *Colour Box* (Lye 1935) was his first attempt at painting colour directly onto film and it was also the first film of this type to be viewed by a wider audience, as it was actually an advertisement for the British Post Office. His film *Free Radicals* (Lye 1958) is one of the earliest examples of abstract animation using percussive rhythmic music, coupling African drumming alongside linear forms which dance to the beat. Lye was also a kinetic sculptor and working artist employed to create artworks for various companies. Fischinger's and Lye's animations could be said to possess musical qualities in the way the objects pulse, move and transform over time. They have taken the ideas used in Kandinsky's paintings and set them into motion.

Seeing Sound and Visual Music

To make her audience aware of the intended relationships between sound and image in her own films, Mary Ellen Bute created a series of works she classed as 'seeing sound'. In these films, Bute visualised music compositions, which in the case of her first colour film, *Synchromy No. 4* (Bute 1938), was Bach's Toccata in D minor. In this film, orange triangles and blue squares interact, move and transform in unison with the musical dynamics and flourish in Bach's composition. Interestingly, Bute also used the term 'visual music' on some of her title slides. This term appeared in print in 1920 (Fry 1920). When discussing the artworks of French post-impressionists, Fry's words echoed the artworks and sentiments of the abstract artists and animators discussed above:

> Now, these artists do not seek to give what can, after all, be but a pale reflex of actual appearance, but to arouse the conviction of a new and definite reality. They do not seek to imitate form, but to create form; not to imitate life, but to find an equivalent for life . . . The logical extreme of such a method would undoubtedly be the attempt to give up all resemblance to natural form, and to create a purely abstract language of form—a visual music.
>
> (Fry 1920, 158)

Further developments in non-objective filmmaking continued throughout the 20th century. From the 1940s brothers John and James Whitney began making abstract films and in the 1950s built and used an analogue computer with a camera controlled by pendulums to animate colourful patterns onto film strip. Their artworks were influenced, among other things, by their interest in Eastern metaphysical beliefs. Films such as *Lapis* (James Whitney 1966) contained evolving circular mandalic imagery akin to some of Jordan Belson's, similarly influenced productions, such as *Allures* (1961) and *Samadhi* (1967) from the same period. Belson had a prolific output and continued making films until a few years before his death in 2011. Stanley Brakhage was another prolific filmmaker who used various techniques such as painting onto film, scratching and baking film to produce many of his works such as *Night Music* (1986). Brakhage was predominantly a silent filmmaker believing the images themselves could create an internal music in the observer.

Hand-drawn Sound and Optical Soundtracks

In some cases existing soundtracks and compositions were used with the above artists' animations but, as already hinted at, novel methods of drawing sounds were also being experimented with during this period. These 'direct' approaches to producing sounds are a type of 'optical sound synthesis' and were utilised during the 1930s simultaneously by German and Russian pioneers. Andrey Smirnov (2005) described four principal techniques:

1 hand-drawn ornamental sound
2 handmade paper sound
3 automated paper sound—the Variophone as a sort of proto-wavetable synthesis
4 spectral analysis, decomposition and re-synthesis technique.

Ornamental sound was created by drawing shapes and curves directly onto cinematic film and, although used by others, it is associated with the work of Oskar Fischinger. Fischinger's drawings were 'ornamental' in that he used graphically appealing shapes that were repeated and modified to create a continuous soundtrack. He described how the shapes and colours influenced the quality of the sound:

> In reference to the general physical properties of drawn sounds, we can note that flat and shallow figures produce soft or distant-sounding tones, while moderate triangulation gives an ordinary volume, and sharply-pointed shapes with deep troughs create the loudest volume. Shades of grey can also play a significant role in drawn music-ornaments. High-contrast definition of the wave form decisively creates the prevalent sound effect, but as long as one places such a 'positive' (well-defined) wave somewhere in the foreground, one can simply overlay other wave patterns simultaneously by using grey shades for the secondary sound effects.
>
> (Fischinger 1932, 1)

Arseny Avraamov also drew ornamental soundtracks in Moscow between 1930 and 1931. Norman McClaren created music in a similar way, with the most notable results of hand-painted soundtracks visible and audible in *Synchromy* (1971). In *Synchromy* the picture and sound are both made from the same hand-painted images in the spirit of 'what you see is what you hear'. The coloured blocks used in this animation are square or rectangular and sound surprisingly like the square wave generated by an electronic oscillator.

Handmade paper sound was essentially the same technique as ornamental sound but the sound shapes were first drawn onto paper and later photographed onto film. This was the approach taken by Rudolf Pfenninger in his early direct sound experiments. Pfenninger developed a technique he called 'Tönende Handschrift' (Sounding Handwriting) (Levin 2003, 53). By analysing the shapes produced by real sounds recorded onto film he was able to recreate his own soundtracks. He drew waveforms onto paper that were like those he had analysed and afterwards transferred to film either by scanning or by recording with a film camera.

The third method, automated paper sound, involved a mechanical device known as the variophone optical synthesiser. The variophone was intended to be used to produce synthesised music rather than to create soundtracks to accompany animations. It comprised a camera and a number of paper disks into which waveforms were cut. The disks were combined and rotated at varying speeds and the moving shapes captured on the camera to produce an optical soundtrack. The variophone was used to help recreate synthesised versions of compositions such as Nikolai Timofeev's 'Waltz', Richard Wagner's 'Ride of the Valkyries' and Franz Liszt's 'Rhapsody No. 6'.

Finally, the spectral analysis techniques of Yankovsky sought to analyse the content of various waveforms and reproduce them graphically so they could later be combined and resynthesised. This was like Pfenninger's method but made through observation of the spectral characteristics of sounds rather than their time-varying qualities.

> Yankovsky firmly believed in the possibility of creating a universal library of sound elements based on the model of Mendeleev's periodic table of elements. Its graphic curves, the 'spectrostandards', were semiotic units which, when combined, formed new hybrid sounds. He also developed several sound processing methods including pitch shifting and time stretching based on the separation of spectral content and formants, resembling recent computer music techniques of cross synthesis and the phase vocoder.
>
> (Smirnov and Liubov 2011, 11)

This taxonomy of sounds allowed Yankovsky to create any sound he desired and also permitted manipulation of the individual sound elements through processes such as time stretching.

Oramics and UPIC

Other methods of optical sound production involved technologies emerging in the second half of the 20th century. Oramics, discussed by Hutton (2003) and invented by Daphne Oram in the 1950s, was a technique where drawings made on 35 mm camera film were processed to manipulate a hardware sound synthesiser. The Oramics machine shone light over the moving film and the light's intensity was modulated by the drawings as it passed through. The light modulation was measured and converted into control information that drove oscillators and other sound synthesis parameters present in the Oramics machine. Oramics therefore used the film drawings as control information rather than producing sound directly through the ornamental type sound production techniques described above. Similarly, Iannis Xenakis, used dedicated apparatus that utilised drawn control and synthesis data, and compositional information. The Unite Polyagogique Informatique du CEMAMu[2] (UPIC), completed in the late 1970s, enabled the user to draw sound waveforms and envelopes on a digitiser tablet. The musician could compose music from these waveforms by drawing the desired pitch information over time across the x-axis on the UPIC tablet. A computer would then reproduce the entire composition. UPIC has since been adapted and revived as the computer program Iannix (Coduys, Jacquemin and Ranc 2021).

Contemporary Visual Music in Academia

Over time visual music as a compositional practice filtered its way into academia and became a subject of intense study and practice-based composition. In 2004, the winter edition of the computer music journal was dedicated to articles and videos discussing contemporary visual music. This was a significant moment in raising my own personal awareness towards developments in audio-visuals and how they could become the focus of further research. Being based in an engineering and digital signal processing department at the time, it was difficult for me to convince colleagues of the merits of the study of this area. Music from a DSP programmer's perspective is treated as a one-dimensional 'vector', images as 2D matrices and video as matrices with a time base! This approach describes artworks in a very quantitative way but says nothing about their qualities. Despite these difficulties in communication there was some receptivity to developing research in composition and gradually I was given space and outlet to continue my creative practise, eventually completing a PhD in 2014 (Payling 2014). Like many of the current academic composers discussed and interviewed in this book, I approach my creative practice with the perspective and skills of a musician. Having filled various roles in conventional four-piece bands during my formative years, I began using computers in music composition in the early 1990s. The Amiga computer I used for sequencing had excellent graphics capabilities for the time, so it was a natural progression to experiment with creating visuals. This, I believe, imparts a particular flavour to my works, as I don't have the dedicated experience and visual adroitness of a true animator or filmmaker. However, with the continuing advancement, accessibility and convergence of audio and visual technologies, it is tempting to utilise these tools creatively. McDonnell (2010) describes the amalgamation of music and video arts as a practice under the sonic arts umbrella and refers to it more specifically as visual music. In this artform there is an exploratory interaction and interplay between the musical elements, often of experimental genres, and the visual, created by animation or video sources. The visual music jointly discussed by Battey and Fischman (2016) first aims to achieve a delicate balance in which all media interact tightly on an egalitarian basis. Second, the organising and structuring of materials in film/animation generally follow an external narrative; visual music, however, often does not, even where there is an implied narrative. Instead, its organisational logic is ultimately dependent on the articulation of materials in time space according to their intrinsic attributes (e.g., shape or colour in the case of images; spectral content and morphology in the case of sounds).

Video Concrète and Video Synthesis

In his creative practice, Bret Battey uses digital animation techniques, whereas Rajmil Fischman uses video footage and other materials. Although these are different technological approaches, the composers have similar motivations, and their compositional outputs can be described as visual music. In this genre it is the relationships between sound and image which are key, not the material properties of either medium. Some comparison of these different approaches is provided here.

Electroacoustic Movies and Video Concrète

Writing about the evolution of his own compositional practice, Joseph Hyde (2012) discussed video concrète, where video clips are treated as entities analogous to sound objects. In this process compositions are created by working with the transformation and development of captured video footage similar to how musique concrète composers treat sound recordings. The results of this approach can be seen in *Vanishing Point* (Hyde 2010). At the time, Hyde described these techniques as being quite rare, but a growing number of artists are creating exceptional compositions via similar methods. Nick Cope, for example, refers to his collaborative compositional works with Tim Howle as 'electroacoustic movies'. Cope's video scratch methodologies align concordantly with Howle's electroacoustic compositions in works such as *Globus Hystericus* (Cope and Howle 2013). Other artists relevant in this context are Inés Wickmann and Francis Dhomont, who have an ongoing collaboration, combining electroacoustic compositions with digital video in pieces such as *Le Silence du Léthé* (2016); João Pedro Oliveira's electroacoustic music inspired videos have become ever more sophisticated and accomplished, as evidenced by *Neshamah* (2018).

Animated Visual Music

Another popular technique used in visual music composition is abstract animation. This follows the trajectory of the pioneering artists discussed earlier who worked with direct animation techniques using film stock. In the mid-20th century, John Whitney propelled animation into the digital realm with his early computer-generated motion graphics. Whitney's geometric vector graphics created harmonious visual patterns observable in his *Matrix I* and *Matrix III* films (1972; 1971) and were an inspiration to my own compositions. In more general terms, animation makes it possible for the artist to be very precise in shape, colour, transformations, motion and other parameters to create any desirable relationship between sound and moving image. This correlates with sound synthesis in music, where the composer has very precise shaping capabilities over each sonic parameter. Also compare this correlation with the one shared between found sound and video footage. Artists currently active in visual music animation include Bret Batty, with his piece *Estuaries 3* (2018). Similarly, Jean Piché is a composer who meshes moving images and music in a new hybrid form he calls 'videomusic'. As an electroacoustic music composer, Piché has produced many multimedia pieces combining video, and more recently animation, with sound in pieces such as *Horizons [Fractured, Folded, Revealed]* (2015).

Visual Music Hybrids

Although the previous discussion segregates visual music according to its production techniques, this is only one way of categorising works. Compositions could also be considered with respect to their use of abstract or representational imagery with abstracted imagery being the prevalent style. There are also hybrid compositions such as *Open Circuits* (Cope and Howle 2003), in which animated Winamp visualisations are combined with video footage. This piece also contains both abstract and real-world representational imagery.

Contemporary Composers and Artists

The preceding discussion provides a glimpse into current creative practice in visual music. It is a rich and diverse field with global reach and subtle variants. Conferences and concerts exist solely for the screening of visual music and many others have audio-visual concerts alongside other artforms, typically music. The annual NoiseFloor Festival[3] has taken place at Staffordshire University since 2010 and I have been involved in its organisation and curation since its inception. It has screened many audio-visual and visual music compositions from a wide variety of composers and curating the audio-visual section of the festival has given me the opportunity to further develop an awareness of current trends and develop a sense of compositional forms and themes that are used. The experience of interacting with the audience and receiving feedback on the screenings has also been useful in informing the direction of future concerts and technical and aesthetic decisions when creating my own work. To bring this discussion to the present day the artists introduced below are a small selection of currently active composers who produce a wide variety of audio-visual and visual music media and had the listed works screened at NoiseFloor.

Francesc Martí is a sound and digital media artist currently living in the UK. His compositions Speech 1 and Speech 2 (Marti 2014a; 2014b) use granular audio and video processing to create rhythmic and glitchy works utilising archive video footage.

Claudia Robles-Angel is a new media and sound artist currently living in Cologne, Germany. She creates audio-visual fixed media compositions, performances and installations. *HINEIN (inwards)* (Robles-Angel 2016) is a gradually evolving visual landscape made from close-ups of natural microstructure and a sonic soundscape of processed found sound recordings.

Hiromi Ishii creates multichannel acousmatic compositions, and Visual Music for which she composes both music and visuals. Her piece *Aquatic* (Ishii 2016) uses heavily treated photographs of fish to create synthetic underwater worlds. The music was created from the recording of whales' voices and further reinforces the theme of the composition. See the interview with Ishii in Chapter 2.

Timothy Moyers Jr is an electroacoustic composer, sound designer and audio-visual artist from Chicago. His piece *Golden Cuttlefish* (Moyers 2020) explores the relationship between the organic and the abstract. A digital ecosystem is created exploring this juxtaposition in both the sonic and visual worlds. Abstract imagery is controlled by organic motion. Organic sound environments coexist with abstract sonic events. The organic flow of musical form and time is complemented by the fluid motion of the video.

Nicola Fumo Frattegiani's research deals with electroacoustic music, soundtracks of images, video, sound theatre and sound exhibition. *Gusseisen* (Frattegiani 2019) is an atmospheric piece created from film footage carefully edited and treated to create cohesion with the dynamic and evocative electroacoustic composition.

Julian Scordato is a composer, sound artist and music technologist. He currently works as Professor of Music Informatics at the Conservatory of Salerno, Italy. *Engi* (Scordato 2017) is an audiovisual work based on a sonification of stellar data related to the north polar constellations. Sound parameters are represented graphically and defined by certain observation data as well as the physical characteristics of stars. Starting from these simple elements for sound generation, this work assumes a certain complexity through the interaction between electronic sounds in a feedback network capable of processing them both diachronically and synchronically.

Additional Influences on Visual Music

Alongside the history of visual music artists and techniques there are several related subjects which have influenced its development.

Sonification

Sonification is a technique that uses 'audio to convey information' (Bonebright et al. 1999) about the data being sonified. Its related discipline of auditory display uses sound to convey information in situations where such means are advantageous. The Geiger counter, for example, emits a granular sound whose density is proportional to the level of radiation. It can also be compared with Chion's causal listening mode (Chion 1994), where a sound is auditioned to gather information about its cause. An example of sonification is the 'Tweetscape' (Hermann et al. 2012) experiment. In this study, Twitter posts were mapped to audio recordings of typewriters and followers' replies were mapped to samples of whispers. The resultant Tweetscapes permitted listeners to determine how many followers the tweet authors had and the number of replies each post received simply by listening to the sonification.

A number of sonification studies have also been undertaken to create sound from visual data. First, one system was devised (Giannakis 2001) whereby small square picture mosaics could be linked together to create combined tones and timbres. This was an experimental synthesis technique that used visual metaphors and interpreted them in sound. The studies from Meijer (1992) and Yeo and Berger (2005) both used pixel scanning techniques to drive synthesis programmes that gave information about pictures. In the case of Meijer's vOICe (Meijer 1992) system, the intention was to aid blind people to 'see' the environment purely with sound. Another experiment (Margounakis and Politis 2006) used 'Chromatic Synthesis' to convert the colour information in a picture to pitch and melody. An image was scanned left to right and top to bottom, as one would read a book, and coloured 'blocks' with contiguous pixels of the same colour were converted into different musical melodies. The left-right scanning technique is quite common and will be discussed further in Chapter 6. It is a successful mechanism for determining various qualities of still or moving images.

Although sonification is not primarily intended to produce music, there have been attempts to assign greater importance to the aesthetic qualities of sonified sound. One of the first sonification concerts took place in Sydney in 2004 and was titled 'Listening to the Mind Listening', curated by Stephen Barrass (2004). Ten pieces were performed, each created from the same data that was taken from a set of electroencephalograph (EEG) traces of a person listening to a piece of music. Composers were invited to use the recorded EEG data to produce music that would be performed in a concert setting. An example of the type of music performed is my own piece, *Listen (Awakening)* (Payling 2004), which used the data to drive a number of different synthesis techniques including formant synthesis, the results of which were mixed and panned to a large array of spatialised monitor speakers. Subsequent iterations of the International Community for Auditory Display (ICAD) conference have included concerts of sonifications of different data sets. The audiovisual composition, *Patterns of Organic Energy* by Sylvia Pengilly (2004), also used sonification techniques to produce sound from video information. Pengilly described how the piece 'is a music/video work in which the sounds used for the music were derived directly from the keyframes of the video, thus creating an intimate link between video and audio' (Pengilly 2004, 118).

Cymatics

Cymatics is a term proposed by Hans Jenny when discussing the study of periodicity and wave vibrations. 'If a name is required for this field of research it might be called cymatics (to kyma, the wave; ta kymatika, matters pertaining to waves, wave matters). This underlines that we are not dealing with vibratory phenomena in the narrow sense, but rather with the effects of vibrations' (Jenny 1967, 20). Jenny devised an experimental method to make sound visible by creating vibrations in physical materials such as glycerine, sand on steel plates and powder on flexible diaphragms. By exciting these physical media with an oscillator or other source, various sonorous figures are created. The specific type of pattern created is dependent on the physical medium and the stimulus used. Jenny studied these effects meticulously, with many audio frequencies applied to different media, and some of the visible results are works of art in their own right. Jenny's work was at least in part inspired by the work of Ernst Chladni. Chladni placed loose sand on a fixed metal plate and made the plate resonate by bowing its edge with a violin bow. As the plate vibrated the sand settled in the nodal lines where there is no vibration on the plate creating 'Chladni figures' (Chladni 1787).

More recent implementations of cymatics use a loudspeaker attached directly to the physical medium. One project (Blore 2021), for example, attached a petri dish to a loudspeaker cone. The dish was filled with various liquids and adjustable lighting created impressive colourful patterns when the speaker was driven by a signal generator. Pure sine tones are more suitable for this type of experiment than complex waveforms such as music. Additional effects can be created by using different physical media. For *Cymatics* (Stanford 2014), a variety of techniques and apparatus were used such as Chladni plates, a hose pipe, ferro fluids, a Ruben's tube and a Tesla coil, all as part of a dynamic performance captured on video. Various instruments caused modulation in each of these media creating a rich visual spectacle. If a non-Newtonian fluid (e.g., corn starch in water) is used in a petri dish attached to a speaker, the results become even more intriguing. The fluid appears to come to life and dance to the sound. Seek out some video examples and prepare to be amazed.

Synaesthesia

Another topic related to sound and vision is the 'condition' known as synaesthesia. Synaesthesia (syn for united, joined and aesthesia—sensation, or union of the senses) occurs when:

> stimulation of one sensory modality automatically triggers a perception in a second modality, in the absence of any direct stimulation to this second modality.
>
> (Harrison and Baron-Cohen 1996, 3)

An example is when a person hears or reads a word, they perceive colours that they associate with its vowels and consonants. The word 'associate' is used cautiously here as the cross-modal perception is not a reasoned one, it is an innate predisposition that the synaesthete has no control over. Since there are five human senses, there are several possible connections between them. For example, a smell could be perceived by the sense of touch. As well as the link between senses there are many other experiences that can cause a response in different senses. Interestingly in this context, however, is that the most common manifestation of synaesthesia has been shown (Harrison and Baron-Cohen 1996) to be coloured-hearing, also known as chromesthesia. This is where the person perceives sounds as colours or vice versa. Although this appears to validate Newton's and others colour-sound correspondences, the associations cannot be generalised between all people with the same cross-modalities. For example one synaesthete may perceive the colour yellow as one sound, and a second synaesthete as another. An example (Motluk 1996) of two synaesthetes discussing their individual perceptions highlights these differences . . .

> If you asked me the colour of a 'horse', I'd rather think of an actual brown horse . . . H as a letter is dark red. But it doesn't affect a word like 'horse'. Well I point out it does for me. 'Horse' is orangey, much like the letter H. The word 'milk' is green because M is green; 'water' is a yellow word, 'bread' blue, 'olive' white and 'snow' red'.
>
> (Motluk 1996, 277)

One theory of how synaesthesia is developed is that the link between the corresponding stimuli could be physiological or emotional. For example, when two stimuli excite two different senses but produce the same physiological response, those stimuli could become linked. A sudden change in pitch or loudness in music could create a physiological reaction similar to one experienced when exposed to flashing colours, for example. More subtle stimuli and reactions could also be experienced and it is possible that many people have some degree of developmental synaesthesia. A greater prevalence of synaesthetic response was also demonstrated amongst children (Harrison and Baron-Cohen 1996), indicating that it is a trait many people share but gradually lose as the learned, culturally ingrained, associations become more dominant. Some of the artists and musicians described in the preceding paragraphs either claimed to have had synaesthesia or exhibited traits that could class them as synaesthetes. It could also be speculated that artistic inspiration can come from many sources that interact to produce ideas and insight.

A recent insight sheds further light on synaesthesia from a neuroscience perspective. The human brain is constantly adapting itself to inputs and new sensations by virtue of its neuroplasticity. When younger, the brain is more adaptable but its propensity for change gradually slows down with age. It is this adaptation and 'configuration' of one's brain that makes every individual unique. In one study, undertaken to learn more about synaesthesia, it was shown that a certain group of individuals shared the same associations between letters of the alphabet and different colours. For example, in the study group everyone associated the letter A with red, B with orange, C with yellow and so on. Demographic data showed that all the participants were born between the late 1960s and late 1980s and it turned out that they had all been exposed to a Fisher-Price magnet set produced between 1971 and 1990 which was on fridge doors across America. Eagleman (2020) therefore describes synaesthesia as a type of 'sticky plasticity'. Instead of the normal process where many different associations average out over time, one association becomes fixed during a specific period of development, creating a synaesthetic response whenever either stimulus is received.

Summary

This historical overview is a brief introduction to a range of artists and influences relevant to visual music. The intention is to show where parallels between sound and image in the arts and sciences exist. A whole chapter could be devoted to each of these artists and subjects, but this summary will hopefully allow rapid identification of individuals and techniques of interest which can then be explored more deeply. There are many resources for further study. For visual music-related studies Aimee Mollaghan's text (2015) gives an excellent insight into early visual music pioneers. William Motriz's Optical Poetry (Moritz 2004) is a thorough discussion of the life and works of Oskar Fischinger. The colourful book *Visual Music: Synaesthesia in Art and Music Since 1900* (Mattis 2005) contains many historical perspectives and a historical timeline. Also, search out opportunities to view some of the films and videos mentioned. Some of these are available online, although many are of poor quality when compared to screenings of film footage or high-definition video. Other works can be purchased on DVD or accessed through paid streaming services from the Center for Visual Music (centerforvisualmusic.org 2010). Most of Norman McLaren's works are hosted in high resolution with open access on the National Film Board of Canada's website (Canada 2021). Much better than viewing through online streaming are events hosted by museums, galleries and universities. Screenings of historical films often take place alongside new compositions at various international conferences with high-quality audio-visual playback.

Notes

1 Lumia will be discussed in relation to composition in Chapter 4.
2 CEMAMu, the Centre d'Etudes de Mathématique et Automatique Musicales. Xenakis' arts and science research centre in France, later renamed CCMIX (Centre for the Composition of Music Iannis Xenakis).
3 See www.noisefloor.org.uk/ for further information about the NoiseFloor Festival.

References

Barrass, Stephen. 2004. 'Listening to the Mind Listening. Concert of Sonifications at the Sydney Opera House'. Music Concert, Sydney Opera House. www.icad.org/websiteV2.0/Conferences/ICAD2004/concert.htm.
Battey, Bret. 2018. *Estuaries 3*. https://vimeo.com/264837797. Audiovisual Composition.
Battey, Bret and Rajmil Fischman. 2016. 'Convergence of Time and Space: The Practice of Visual Music from an Electroacoustic Music Perspective'. *The Oxford Handbook of Sound and Image in Western Art*, August. https://doi.org/10.1093/oxfordhb/97801998 41547.013.002.
Belson, Jordan. 1961. *Allures*. 16 mm. Short.
———. 1967. *Samadhi*. Animation. Short.
Bishop, Bainbridge. 1893. *The Harmony of Light*. New Russia, Essex County, NY: The De Vinne Press.
Blore, Dan. 2021. 'Cymatics.Org'. Cyamtics.Org. 2021. http://cymatics.org/.
Bonebright, T., P. Cook, J. Flowers, N. Miner, J. Neuhoff, R. Bargar, S. Barrass, J. Berger, G. Evreinov, and W. T. Fitch, M. Grohn, S. Handel, H.G. Kaper, H. Levkowitz, S.K. Lodha, B.G. Shinn-Cunningham, M. Simoni and S. Tipei. 1999. 'Sonification Report: Status of the Field and Research Agenda Prepared for the National Science Foundation by members of the International Community for Auditory Display'.
Brakhage, Stan. 1986. *Night Music*. Short.
Bute, Mary Ellen. 1938. *Synchromy No. 4: Escape*. Colour Animation Short.
Canada, National Film Board of. 2021. 'NFB Films Directed by Norman McLaren'. National Film Board of Canada. 2021. www.nfb.ca/directors/norman-mclaren/.
Castel, Louis-Bertrand. 1740. *L'optique Des Couleurs, Fondée Sur Les Simples Observations, & Tournée Surtout À La Pratique De La Peinture, De La Teinture & Des Autres Arts Coloristes . . .* Paris: Chez Briasson.
centerforvisualmusic.org. 2010. 'Center for Visual Music'. Vimeo.com. 2010. https://vimeo.com/channels/124018.
Chion, Michel. 1994. *Audio-Vision: Sound on Screen*. New York: Columbia University Press.

Chladni, Ernst Florens Friedrich. 1787. *Entdeckungen Uber Die Theorie Des Klanges*. Leipzig: Bey Weidmanns Erben und Reich.

Coduys, Thierry, Guillaume Jacquemin and Matthieu Ranc. 2021. 'IanniX'. Iannix. 2021. www.iannix.org/en/.

Collopy, Fred. 2000. 'Color, Form, and Motion: Dimensions of a Musical Art of Light'. *Leonardo* 33 (5): 355–60.

———. 2009. 'Playing (With) Color'. *Glimpse | the Art + Science of Seeing* 2 (3): 62–67.

Cope, Nick and Tim Howle. 2003. *Open Circuits*. Electroacoustic Movie. https://vimeo.com/633346.

———. 2013. *Globus Hystericus*. Electroacoustic Movie. https://vimeo.com/226161880.

Eagleman, David. 2020. *Livewired: The Inside Story of the Ever-Changing Brain*. Illustrated edition. New York: Pantheon Books.

Eggeling, Viking. 1924. *Symphonie Diagonale*. Animation.

Fischinger, Oskar. 1932. 'Sounding Ornaments'. *Deutsche Allgemeine Zeitung*, 1932. www.oskarfischinger.org/Sounding.htm.

———. 1937. *An Optical Poem*. Animation, Short.

———. 1947. 'Fischinger—My Statements'. Oskarfischinger.Org. 1947. www.oskarfischinger.org/MyStatements.htm.

Franssen, Maarten. 1991. 'The Ocular Harpsichord of Louis-Bertrand Castel'. *Tractrix* 3: 15–77.

Frattegiani, Nicola Fumo. 2019. *Gusseisen*. Audiovisual Composition. https://vimeo.com/351652309.

Fry, Roger. 1920. *Vision and Design*. London: Chatto and Windus.

Gerstner, Karl. 1981. *Spirit of Colour: Art of Karl Gerstner*. Cambridge, MA: MIT Press.

Giannakis, Konstantinos. 2001. 'Sound Mosaics a Graphical User Interface for Sound Synthesis Based on Auditory-Visual Associations'. PhD dissertation. School of Computing Science: Middlesex University.

Harrison, John E. and Simon Baron-Cohen. 1996. 'Synaesthesia: An Introduction'. In *Synaesthesia: Classic and Contemporary Readings*, 269–77. Oxford, UK: Wiley-Blackwell.

Hermann, Thomas, Anselm Venezian Nehls, Florian Eitel, Tarik Barri and Marcus Gammel. 2012. 'Tweetscapes-Real-Time Sonification of Twitter Data Streams for Radio Broadcasting'. In *Proceedings of the 18th International Conference on Auditory Display*. Atlanta, Georgia, USA.

Humberstone, Lloyd. 1991. 'Timbre Noun—Definition in the Collins English Dictionary'. In *Collins English Dictionary—Third Edition*, 3rd ed., 1613. Glasgow: Harper Collins.

Hutton, Jo. 2003. 'Daphne Oram: Innovator, Writer and Composer'. *Organised Sound* 8 (1): 49–56.

Hyde, Joseph. 2010. *Vanishing.Point*. Audiovisual Composition. https://vimeo.com/10216134.

———. 2012. 'Musique Concrète Thinking in Visual Music Practice: Audiovisual Silence and Noise, Reduced Listening and Visual Suspension'. *Organised Sound* 17 (02): 170–78.

———. 2022. 'Cross-Modal Theories of Sound and Image'. In *Live Visuals. History, Theory, Practice*, 135–63. London: Routledge. https://doi.org/10.4324/9781003282396-9.

Ishii, Hiromi. 2016. *Aquatic*. Visual Music. https://vimeo.com/156292151.

Jenny, Hans. 1967. *Cymatics: A Study of Wave Phenomena and Vibration*. Newmarket, NH: Macromedia Press.

Jones, Tom Douglas. 1972. *Art of Light and Colour*. New York: Van Nostrand Reinhold.

Kandinsky, Wassily. 1914. *Concerning the Spiritual in Art*. Dover Publications Inc. 1997. New York: George Wittenborn Inc.

————. 1923. *Composition VIII*. Oil on Canvas. www.wassilykandinsky.net/work-50.php.

————. 1939. *Composition X*. Oil on Canvas. www.wassilykandinsky.net/work-62.php.

Lavignac, Albert. 1899. *Music and Musicians*. New York: Henry Holt and Company.

Levin, Thoma Y. 2003. '"Tones from out of Nowhere": Rudolph Pfenninger and the Archaeology of Synthetic Sound'. *Grey Room*, 12: 32–79.

Lye, Len. 1935. *A Colour Box*. Hand Painted Animation.

————. 1958. *Free Radicals*. Black and White Animation.

Margounakis, Dimitrios, and Dionysios Politis. 2006. 'Converting Images to Music Using Their Colour Properties'. *Proceedings of the 12th International Conference on Auditory Display*, June.

Marti, Francesc. 2014a. *Speech 1*. Audiovisual Composition. https://vimeo.com/106805757.

————. 2014b. *Speech 2*. Audiovisual Composition. https://vimeo.com/119713106.

Mattis, Olivia, ed. 2005. *Visual Music: Synaesthesia in Art and Music Since 1900*. Illustrated edition. New York: Thames & Hudson.

McDonnell, Maura. 2010. 'Visual Music—A Composition of the "Things Themselves"'. In *Sounding Out Conference*. Vol. 5. Bournemouth: Bournemouth University. www.academia.edu/525221/Visual_Music_-_A_Composition_Of_The_Things_Themselves.

McLaren, Norman. 1971. *Synchromy*. Hand-drawn Animation.

Meijer, Peter. 1992. 'An Experimental System for Auditory Image Representations'. *IEEE Transactions on Biomedical Engineering* 39 (2): 112–21. https://doi.org/10.1109/10.121642.

Mollaghan, Aimee. 2015. *The Visual Music Film*. 1st ed. 2015 edition. London: Palgrave Macmillan.

Moritz, William. 2004. *Optical Poetry: The Life and Work of Oskar Fischinger*. Bloomington: Indiana University Press.

Motluk, Alison. 1996. 'Two Synaesthetes Talking Colour'. In *Synaesthesia: Classic and Contemporary Readings*, edited by John E. Harrison and Simon Baron-Cohen, 269–77. Oxford, UK: Wiley-Blackwell.

Moyers, Timothy. 2020. *Golden Cuttlefish*. CGI Animation. https://vimeo.com/427785668.

Newton, Sir Isaac. 1704. *Opticks: Or a Treatise of the Reflections, Inflections, and Colours of Light*. London: The Prince's Arms in St Paul Church Yard.

Oliveira, João Pedro. 2018. *Neshamah*. Electroacoustic Video. https://vimeo.com/255169571.

Payling, Dave. 2004. 'Listen (Awakening): A Composition with Auditory Display'. *ICAD 2004: The 10th Meeting of the International Conference on Auditory Display, Sydney, Australia, 6–9 July 2004, Proceedings*.

Payling, David. 2014. 'Visual Music Composition with Electronic Sound and Video'. Stafford, UK: Staffordshire University. http://eprints.staffs.ac.uk/2047/.

Pengilly, Sylvia. 2004. *Patterns of Organic Energy*. Audio Visual.

Piché, Jean. 2015. *Horizons [Fractured, Folded, Revealed]*. Videomusic. Montreal. https://vimeo.com/128936189.

Richter, Hans. 1925. *Rhythmus 21*. Animation, Short.

Rimington, Alexander Wallace. 1895. Method and Means or Apparatus for Producing Colour Effects. London GB 189324814, filed 24 September 1894, and issued 1895.

Robles-Angel, Claudia. 2016. *HINEIN (Inwards)*. Audiovisual Composition. https://vimeo.com/200334611.

Ruttmann, Walter. 1921. *Lichtspiel Opus 1*. Animation, Short, Music. Ruttmann-Film.

Scholes, Percy A. 1950. *The Oxford Companion to Music*. 8th ed. London: Oxford University Press.

Scordato, Julian. 2017. *Engi*. Audiovisual Composition. www.julianscordato.com/works. html#engi.

Smirnov, Andrey. 2005. 'Sound out of Paper'. Andrey Smirnov. 2005. http://asmir.info/ gsound1.htm.

Smirnov, Andrey and Pchelkina Liubov. 2011. 'Russian Pioneers of Sound Art in the 1920s'. Madrid. http://asmir.info/articles/Article_Madrid_2011.pdf.

Stanford, Nigel. 2014. *Cymatics—Science Vs. Music*. https://NigelStanford.com/Cymatics/.

Stein, Donna, M. 1971. 'Thomas Wilfred: Lumia'. *Press Preview*, 8 September 1971.

Whitney, James. 1966. *Lapis*. Colour, Mono sound. Computer Animation.

Whitney, John. 1971. *Matrix I*. Animated Short. www.imdb.com/title/tt5968306/.

———. 1972. *Matrix III*. Animated Short. www.imdb.com/title/tt2600366/.

Wickmann, Inés and Francis Dhomont. 2016. *Le Silence du Lethe*. Electroacoustic Video. https://vimeo.com/170500471.

Wilfred, Thomas. 1947. 'Light and the Artist'. *The Journal of Aesthetics and Art Criticism* 5 (4): 247–55.

Yeo, Woon Seung and Jonathan Berger. 2005. 'A Framework for Designing Image Sonification Methods'. *Proceedings of ICAD 05-Eleventh Meeting of the International Conference on Auditory Display*, June.

2 Communicate

Electronic Visual Music Conversations

Louise Harris

Louise is an audiovisual composer and Professor of Audiovisual Composition at the University of Glasgow. Her work has been screened and performed at conferences and events both internationally and locally. She has published widely on her creative practice in a number of journal articles and chapters and her first book, *Composing Audiovisually*, was published recently (2021). Louise was a source of inspiration and guidance to me when completing my own PhD.

Dave: *What inspired you to begin creating audio-visual works? Who or what are your main influences and why?*

Louise: I think like a lot of kids who are interested in composing, and who also like films, when I was younger I wanted to be a film composer. I became fascinated with the relationship between sound and image at the age of about 2 or 3 – my mum and dad took me to see the rerelease of Disney's *Pinocchio* at the cinema and I still remember it now. The music, the sound, the drama—everything was so overwhelming, so grand and immersive. So, from that point, probably without knowing it, I was fascinated with film and how music/sound fit within it. From the age of about 4 I was fascinated by music in its own right—I heard someone play the flute at a school assembly and just loved the sound it made. I had to wait until I was 10 to learn to play through school (which felt like a long time!) but from then I was just making music all the time—orchestras, choirs, wind bands, flute choirs and so on. By the time I went to study music at university I was playing/singing in groups at least six or seven times a week. At uni, I was able to start exploring composition but felt like I really struggled to 'get at' the sound I wanted, working with purely acoustic means—I had no idea about the possibilities of working with recorded sound or technology of any kind, and it wasn't really a thing my university 'did' at that point, so I had no real means of encountering it. It wasn't until I was about a year into my PhD (in composition but specialising in acoustic and particularly vocal music) that I made my first work with recorded sound—a short clip of a slide projector from a gallery in Sheffield. Once I'd made the sound work, I had a clear sense of a visual accompaniment I wanted to go with it, so I asked friends and colleagues to help me figure out how to make the visual materials I wanted. This was my first audiovisual

DOI: 10.4324/b23058-3

work—*Site*—and was the first work that really clicked for me; the first piece I made that made complete compositional sense in my brain. At this point, I was a total newbie when it came to audiovisual practices as a whole, so people started to make suggestions of things to watch, read, listen to . . . I remember the first time I saw McLaren's *Synchromy* and I just felt like I wanted to cry; everything made total sense in how the pieces of the compositional puzzle worked together, it was like I'd found the thing I'd been missing. From there I began to explore visual music in more depth, while also trying to fill in my knowledge gaps on the electronic/electroacoustic music side, which I'd never really explored up to that point. It was a really steep learning curve, but also as I came to it through my own exploration it felt like quite an organic development—one in which I could explore, develop and experiment as I wanted to as opposed to being required to look at/listen to/read particular things. It's an approach I try to bring into my teaching practice now, when getting students to explore new art fields—making suggestions for starting points, but not being prescriptive about what they should/shouldn't encounter.

Which one of your compositions or performances would you recommend to someone new to your work and why?

I think *fuzee* (Harris 2011) and *alocas* (Harris 2017) are both good starting points—there's a really clear relationship between sound and image, without it being an obvious or overly direct 1:1 mapping. I think it's also clear to see how my practice has developed between those two works (and I also think they're just a couple of my best and most engaging ones).

Did you start as a single discipline artist, for example, musician/video artist and how has that coloured your experience working with multiple media? If you always worked with multiple media, how did you simultaneously develop skills and interests in both media?

I've answered this a bit above, but very much so—I trained as a musician, first as a classical flautist before beginning to explore composition more at university, and later exploring audiovisual work during my PhD and afterwards. I think because my educational background is very much concentrated on the musical side I've always felt like a wee bit of a fraud with regard to the visual elements of my work—I have a really developed interest in cinema and visual art, but because I don't have an educational background in that area it has sometimes felt like the less developed aspect of my practice.

How do you manage the relationships between sound and image? Are they equally important, does one lead the other and/or does this change with each project?

They're always equally important—that's central to my whole process, and to how I think and talk about it. Even if they're not composed at the same time, or conceived of at the same time, if I know I'm making a work that is audiovisual then I think about the composition as a fully audiovisual process—neither sound nor image are more important. Often, one will suggest the other if I make them in stages—so while I'm recording/singing/playing/manipulating sound, often colours or forms/shapes/movement will seem to suggest themselves, and vice versa.

What is your approach to live audiovisual performance and how does it differ to composition? For your live pieces, why do you choose to perform them rather than compose fixed media?

For me, the nature of the relationship between the sound and image is as equal and as important in my live works. The main thing about performing is the liveness of it—the decision-making in real-time and the improvisational element. Fundamentally, I think I just missed performing—having done almost nothing else during the formative years of my life, the older I got and the more my practice developed, the less I tended to perform, in any form. Through performing my pieces live, I get back some of that adrenaline rush that I missed from performing historically. I also like to build systems for performance that have a certain amount of agency and unpredictability, which makes it feel like a more collaborative performance than just me by myself.

What are the main tools and technologies that you use for sound and image and do they change dependent on project?

Usually max/jitter for live work and max/jitter/processing for fixed media work. I sometimes do things in After Effects if I'm struggling to achieve what I want else-where, and have flirted with things like touch designer, but mostly I use max/jitter because that's also a big part of what I teach so I spend half my life playing with it.

I would like to include more details of two of your performances, ic2 (intervention:coaction) and NoisyMass, but please discuss only one of these if both will involve too much detail. Could you provide information about the approaches you took both artistically and technically with these? Did ic2 influence or inform NoisyMass in any way or are they completely stand-alone and unrelated works?

ic2 (Harris 2014) was one of my first explorations in taking my practice into live performance. It evolved out of working with Kingston University Digital Arts Collective (KUDAC) while I was teaching there and was a way of challenging myself to take my fixed media practice into a live domain. At that point, I was using predominantly puredata (pd) and processing, with open sound control (OSC) to communicate between the two. It was initially a way of pushing myself to think about how to make the things I want to in real-time, but developed over time to become a part of my research and thinking about the way I make work.

NoisyMass (Harris 2018) was something a bit different. I had been wanting to make a live performance version of the children's memory game, 'Simon', for a while and as I was doing a bit of experimenting with simple live electronics at the time decided to build my own device to do so. The idea was to take the basic premise of Simon—remembering an incrementally extending series of lights and accompanying tones—but extend it to a grid of 3x3 squares of coloured lights/ sounds and build in a series of possible behaviours to make it a little less predictable. So mostly it'll just play a gradually extending series of lights/tones that I have to match, but every now and then it'll interrupt me or change the sequence midway through, or drop out completely for a while before restarting in a completely new sequence. Equally, while Simon is just a series of simple tones from a total of four, every time the sequence triggers it can reset the series of associated sound samples, so it doesn't always sound the same. It's really fun to play with, if slightly nerve-wracking in performance because it can in theory go on for hours (or as long as I can remember the sequence, as I only get three tries each time!) but if I make a mess of it or it decides to intervene, things can go in interesting directions. The plan was ultimately to make a giant version of the work, one in which the individual lights/buttons would be distributed around a large stage area and I would have to run between them to complete the sequence. I haven't quite got round to that yet, but it's on my to-do list! It's an interesting work, because in many ways it's not 'audiovisual' in the way a lot of my others are, but I think it speaks instead to my interest in the relationship between the two—I couldn't remember the extending sequence anywhere near as well if it was just lights or just sounds. Combining the two changes the way I respond, which pretty much sums up how I think about audiovisual relationships in general.

What are your future plans for composition and/or performances?

At the moment, I'm working on a series of pieces based around data audiovisualisation as part of a research fellowship I've been lucky enough to get some funding for. I'm using data of a range of types and from a range of sources to see whether a) it's possible to gain new insight into the data by exploring it audiovisually and b) whether I can use the data as a means of compositional structuring. It's still very early days but some interesting things are emerging. I'm anticipating a series of fixed media works out of that project, and possibly some ongoing collaborations with some of the scientists I've been working with. So we'll see how it develops.

Performance-wise, I'm not actually working on anything just now! I've instead been making some works that are to be interacted with but by people other than me, such as *filigree traces* (Harris 2021), which is intended to encourage participates to think about the audiovisual experience as they explore natural environments. That work was the first I've done that didn't involve screens, speakers cand so on— in fact, no electricity at all, everything is people-powered (and consequently just about the easiest install I've ever done!)—and I'm wondering about taking some of those elements into a live performance environment. Something to explore more once the fellowship is complete.

Tim Thompson

Tim is a software engineer, musician and installation artist. He designs and creates interactive audio-visual instruments which he tours at various festivals in the USA, including Burning Man. His creations are accessible and engaging, giving new and experienced users an enjoyable way into experiencing visual music performance.

Dave: *Who are your main influences artistically and creatively?*

Tim: For visual influences, Wassily Kandinsky and Oskar Fischinger are noteworthy. Thomas Wilfred's work is the most inspirational for me, partly because of the beautiful results achieved with Lumia, but mostly because of his work creating interactive visual instruments with which he performed visual concerts, as well as self-running visual installations he created to produce ever-changing visual output in a form suitable for the home. Modern influences include many artists creating interactive works with cameras, projectors and computers. Scott Snibbe's *Deep Walls* (Snibbe 2002) and Golan Levin's *Yellowtail* (Levin 1998) were seminal. After going to Burning Man for the first time in 2002, my work changed in a fundamental way; rather than building things intended for my own creative use, I started building things to foster and enhance creativity for other people in public settings.

How did you become interested in visual music and what inspired you to build the Space Palette? Can you describe the Space Palette, its construction and the types of events that you visit with it?

After developing algorithmic music software for my own use over several decades (1985–2005), I became interested in extending my work into the visual world when GPUs, cameras and depth cameras became more affordable in the early 2000s. LoopyCam (Thompson 2010) was my first interactive visual instrument, attaching a security camera to a USB number pad, creating a handheld controller on a 16-foot cable that could capture, process and mix up to eight video loops in real-time. Next, the iGesture pads from Fingerworks introduced me to the expressive potential of 3D input using your fingers (finger area being the third dimension), and I created an interactive controller (Thompson n.d.) for performing visuals with them. Ever since then I've focused on using 3D input as expressive input to software generating visuals and music simultaneously.

The Microsoft Kinect was the game-changing inexpensive depth camera which enabled 3D input with your hands in mid-air, and in 2011 I deployed a prototype visual music installation/instrument at my camp at Burning Man. In 2012 I returned to Burning Man with the first oval version of the Space Palette (Thompson 2021) which was extremely well enjoyed, and it has been enjoyed at numerous Burning Mans and other festivals over the last decade.

In 2017, I used the Sensel Morph (a multitouch pressure-sensitive pad with incredible sensitivity and dynamic range) to create the first version of the Space Palette Pro, which has the same basic operation as the original Space Palette—four 3D input devices with which you finger paint visuals and musical notes. The difference is that you're finger painting on pads versus finger painting in mid-air. Over the last few years, I've refined the physical design of its cabinet and open-sourced (Thompson [2021] 2022) the CNC data and software for it. The Space Palette Pro consists of a bespoke wood cabinet containing a Windows computer, four Sensel Morphs, a touchscreen for preset control, a monitor for the visual output, and speakers/ headphones for the sound output. The commercial software I use is Resolume (for visuals) and Omnisphere (for sound). The software I've written and open-sourced consists of the touchscreen GUI for presets, a real-time engine which takes the 3D input from the Morphs and sends MIDI/OSC to Omnisphere and Resolume, and a FreeFrame plugin which runs inside Resolume. I occasionally give talks which describe the nerdy details of the Space Palette Pro (Thompson 2020), or my overall exploration of 3D input (Thompson, n.d.).

Both versions of the Space Palette have similar musical and visual behaviour; you are using four 3D inputs, each controlling a separate visual music instrument, all four instruments being mixed to produce the final output. Musically, the X position of your hand or fingers controls the pitch of the notes being played (forced onto a scale to be musical), the Y position controls the quantisation of the notes, and the Z position (depth or pressure) controls the volume, vibrato, or filtering of the notes. Visually, the X and Y positions control the placement of sprites (processed with various filters to be visually interesting), and the Z position controls their size. By controlling both visuals and notes from the same input source—your hands and fingers—they are naturally synchronised. Both versions have a selection of presets to choose from a variety of sounds and visuals.

How do Space Palette performers respond? Does their mastery improve over time, are some people more adept, do you have repeat visitors?

Particularly at Burning Man, when people have an entire week to play with whatever you bring, repeat visitors are extremely common, usually accompanied by their friends. The Kinect-based Space Palette is the most readily engaging, with a large physical presence and large free-space hand motion; you're essentially dancing with your hands inside four 3D 'mousepads in mid-air'. Watching other people discover and play it can be as enjoyable as playing it yourself.

I call my installations 'casual instruments', analogous to 'casual games', where the mechanics of using it are simple and immediately engaging with little barrier to entry. Using the Space Palette is like finger painting, and producing both visuals and notes simultaneously provides plenty of feedback to the players. Each of the four pads (or holes, in the original Space Palette) lets you play a different visual and sound; you are conducting a visual music orchestra where you are playing all of the instruments simultaneously. There is enough room (particularly in the original Space Palette) for multiple people to play together, and this is extremely common. The variety of presets with different visuals and sounds, along with the lack of pre-recorded media, provides an open-ended experience in which each player can produce a completely different and personal result.

What feedback do users give you when using the Space Palette?

The most common statements heard in the first few years were literally 'I want one in my living room' and 'I could stay here all night'. At Burning Man one year, someone walked out of a tent containing four smaller Space Palettes and said

'I never knew I was creative until I walked in there'. At Symbiosis I had a conversation with Sierra (Thompson 2012), who had just finished playing:

Sierra:	Some people who are so, who don't have a connection with music necessarily—they're not a singer, they don't play an instrument . . .
Tim:	. . . Well, that's what this is for, those people . . .
Sierra:	They . . . that . . . Wow! I mean, I'm a singer, I'm a dancer, I play instruments, so I've been blessed to have that connection my whole life, but for those who can't cross that barrier, literally they're crossing that barrier right here (as she stretches her arms through the Space Palette to illustrate).

For people who don't consider themselves creative or musical, 'crossing that barrier' is a definite goal of the Space Palette.

The Space Palette could possibly work with just sound and just visuals, how important is it to have sound and image working together?

It's important because it's so enjoyably creative and even intimate when absolutely everything you're seeing and hearing is directly controlled by you. I occasionally use it in visuals-only mode to accompany other musicians, and it works very well for that because the visuals are completely live (as opposed to using pre-recorded video clips) and you can adapt more freely to whatever the musician is doing. The 'casual instrument' GUI of the Space Palette Pro has a secret mode that allows real-time editing of the visual parameters, as well as looping, enhancing the performance possibilities significantly. I occasionally expose the looping behaviour on the 'casual instrument' GUI, although players new to the instrument and looping in general can get easily confused, so in completely unattended settings I usually hide the looping features.

What other visual music performance systems/performers are you aware of?

I'm sure I'll be forgetting some good systems/performers, but I'll mention a couple of systems I find interesting. Sushi Dragon, a streamer on twitch.tv, has a remarkably flexible visual performance system using handheld Twiddler controllers and many cameras. Eboman has a camera-based performance system called EboSuite (Eboman n.d.). I find the Playmodes screen ensemble (Playmodes 2022) an interesting visual music system. Dan Tepfer has piano-driven visuals directly attached to the notes he's playing, using his own software as well as some other software intended for live visuals driven from MIDI keyboards, for example Keysight (Sweet 2020). I think there are several performers on twitch.tv using similar software, providing visual output directly attached to a live musical performance, and I expect this kind of software to blossom in the future. There are of course a number of terrific platforms that can be used to build such systems: Isadora, Resolume/Wire, TouchDesigner, Max for Live, and so on.

What does the term visual music mean to you and why did/do you choose that term to describe your work?

It's a term that is often applied to pre-recorded work, but it can also be applied to an extremely wide range of possibilities when you include interactive and performative work. I use phrases like 'casual instrument', 'interactive installation', and 'visual music instrument' to describe the interactive, performative, and improvisational contexts for which my work is usually intended.

What future plans do you have for the Space Palette?

There are new stereo depth cameras that work in sunlight (unlike the Kinect), so I'm planning on making use of those so that the original Space Palette can work outdoors during the day. I consider the latest Space Palette Pro (the one I've open-sourced the design for) to be a platform for interface experimentation that I'll be using for the next decade. The current interface is very direct and basic, because of the 'casual instrument' focus, but there is lots of potential for more interesting behaviour—two-handed semantics, drawing figures and then manipulating them, playing small sequences rather than notes, Samchillian and other novel playing approaches, and so on.

João Pedro Oliveira

João holds the Corwin Endowed Chair in Composition for the University of California at Santa Barbara. He studied organ performance, composition and architecture in Lisbon. He completed a PhD in Music at the University of New York at Stony Brook. His music includes opera, orchestral compositions, chamber music, electroacoustic music and experimental video. He has received over 70 international prizes and awards for his works, including three Prizes at Bourges Electroacoustic Music Competition and the prestigious Magisterium Prize and Giga-Hertz Special Award. He taught at Aveiro University (Portugal) and Federal University of Minas Gerais (Brazil). His publications include several articles in journals and a book on 20th century music theory.

Dave: *What inspired you to begin creating audio-visual works? Who or what are your main influences and why?*

João: I studied music and architecture both at the same time. Although I never worked as an architect, the relations between sound and visual arts always were an important part in my creative process. Notions of proportion, space, movement, remained central in the conception and realisation of most of my music compositions. Moving into visuals seemed to be a natural step, but it was taken quite late in my life. On several occasions I worked with visual artists in different projects, at some point in my life (around 2013) I decided to experiment in the audio-visual realm, also because the technology available at that time was already quite powerful and allowed a fairly easy workflow, especially in the graphics part. Since then, I tried to evolve in different directions and research possibilities of interaction between sound and image.

My influences are multiple. Some of my main references when I started working in audio-visual were some of the classic movies of science fiction (*2001: A Space Odyssey*, *Solaris*, *Forbidden Planet*, and others), as well as many of the animation classics (McLaren, Fischinger, Disney). I also knew the work of Dennis Miller and other artists that were working in the area, and for whom I had an enormous admiration. Nowadays I try to search and find different proposals in audio-visual work and, fortunately, there are many new ideas and developments that make this field extremely rich and innovative.

How has your experience as a music composer coloured your experience working with multiple media?

All the experience I had in the area certainly influenced my approach to visuals. I was very interested in musical gesture and texture (as defined by authors such as Hatten or Smalley) and how it could be used to create the overall shape of phrases or sections of pieces, as well as the opposition between tension and relaxation in the musical flow. Adventuring into the moving image realm allowed me to use all that experience and apply it to the relationship between what happens in sound and what happens in image at a certain moment, how both evolve in time, how they can be used as counterpoint, analogy, opposition, and so on. In images with movement, sound can combine to achieve a goal or support the energetic flow of the image; or possibly contradict that movement. Gestural actions in the image and sound (here interpreted as meaningful energetic changes, or movements based on causality) can have a special role, implied by all the semiotic levels they can potentially carry.

Following Chion's ideas, I started looking for possibilities of achieving the 'synchresis' effect that he describes in his texts: 'the spontaneous and irresistible weld produced between a particular auditory phenomenon and visual phenomenon when they occur at the same time'. All of this was supported by researching into the technical possibilities that existed for visuals (understanding the concepts, exploring the software, applying different methods of image synthesis or manipulation, experimentation, analysing results).

How do you manage the relationships between sound and image? Are they equally important, does one lead the other and/or does this change with each project?

Since one of my main approaches to sound and image is gesture related, both media are conceptually important. However, when establishing the relation between them, sometimes one takes the lead. For example, I have sections in pieces where the sound was composed first and image was added afterwards, other sections where image came first and sound followed, and some other sections where the two of them followed a quasi-parallel path in the composition process. While working on a piece, several gesture relationships between sounds and images can be considered: for example, sound gestures can combine synchronously with a gestural image, or sound textures may oppose (or form a counterpoint) a gestural image; textures in image may associate with textures in sound; gestures in image and sound may combine, but not interact synchronously; gestures in sound may be used to define articulations in sections where the image is static or with little movement. Since musical gestures emerge from the combination of various elements including articulation, dynamics, pitch, duration, and so on, and they acquire a special meaning when these elements coordinate to provoke meaningful energetic changes, it is possible that the listener/viewer will understand 'gestural' meanings intuitively. If we accept this idea, then it is possible to find associations between gesture in sound and image that can have multiple connotations or go beyond the simple 'transcription' of one into the other, connecting them in many different and fascinating ways.

The idea of organicity was also an important point of departure. By organicity, I mean a coherently constructed (yet somehow 'natural') connection between image and sound. For that to happen, several problems were taken into consideration:

- If a moving image and sound both project themselves in time, is it possible to construct or develop transformations that also progress in time and can be applied to both, in order to establish a connecting bond? Can these transformations depart initially from the characteristics of one media (structural, formal, perceived, imagined, etc.), and be used as a model for transformations in the other?
- If audio-visuals are concerned with synaesthetic relations, which parameters can we explore in a composition that simultaneously stimulate several senses, in order to achieve the phenomenon of synchresis?
- When working with images and sounds, and trying to provide answers to the questions above, to which extent does intuition play an important role in the construction of sound-image relations? How much theoretical/structural/experiential knowledge or background is supporting any intuition-based decision? In what way can intuitive insights connect with a logical or structural approach to composition and collaborate actively in the process of creating bonds between sound and image?

The above-mentioned concept of gesture partially answers some of these questions, especially when we deal with the idea of movement or motion interpretation in both media. In other words, how can we connect sound 'spectromorphology' and image 'imagemorphology' characteristics in interesting and creative ways.

How do you approach a new composition—with a concept, a technology, a call/ commission . . . ? As things progress, how much do you improvise with the developing artwork and how much do you stick to the original concept/compositional intent?

That changes from piece to piece. Sometimes there is an initial concept, then there is the 'laboratory' period where experiments (we can call them time-extended improvisations) are made, trying to give a visible/audible shape to that concept. Sometimes the results achieved in the experimentation period change or improve the original concept and give shape to the composition in a way that was partially not intended in the initial stages of the creative process. Composing becomes a dialectical process where the results obtained have to be constantly analysed and updated according to the evolution of the piece and the experimentation consequences. My piece, *Things I Have Seen in My Dreams* (Oliveira 2019), was composed in such a way.

Other times, the initial concept is maintained from the beginning to the end of the piece, and all composing actions are determined by that concept. Timings, phrase, gesture construction and other micro-level actions can be changed or adapted, but the overall trajectory of the piece is determined initially. Such is the case of *Tesseract* (Oliveira 2018b), a piece I composed inspired by the possible projections of a hypercube.

What are the main technologies that you use for sound and image and/or do they change dependent on project?

I believe that technology's sole purpose is to support creative thought. It should not have a role of its own, but mainly serve the creative role imagined by the artist. I usually search for the technology that allows the best realisation of a creative idea. I am interested in exploring any program or software, if I find that the possibilities they offer can give shape to the musical or visual ideas I am trying to implement. I have used most of the available free and commercial software (AfterEffects, Final Cut, Cinema 4D, Mandelbulb, Blender, etc.).

Can you say more about the approach you took in creating Neshamah (Oliveira 2018a), *both artistically and technically? What inspired this piece?*

Neshamah was composed in 2016 at the Centro Mexicano para la Música y las Artes Sonoras (Mexico), and Centro de Pesquisa em Música Contemporânea at Federal University of Minas Gerais (Brazil), and at the composer's personal studio. This piece was done with the support of an Ibermúsicas project. It belongs to a cycle of four pieces inspired by representations of the four elements (fire, water, earth, air) in the Old Testament. Neshamah is a Hebrew word that means 'breath', relates to the air element and was inspired by the first reference to this element in the book of Genesis, used as a metaphor for the creation process:

> Then the Lord God formed the man of dust from the ground and breathed into his nostrils the breath of life, and the man became a living creature.
>
> (Genesis 2:7)

The initial proposal for this project was to compose a piece for acousmatic music and dance, to be performed live, with the collaboration of the choreographer and dancer Rosario Romero. Based on the music, Romero created all the choreographic movements and footage that were used to compose the live version of the piece and the final video. The process of composition itself went through several phases, and was always focused on the collaboration, discussion and exchange of ideas between me and the choreographer/performer. Several music sections of the work were composed and presented to the dancer as a stimulus for the choreographic movements or, in some other instances, the dancer proposed specific body positions or movements. With the progress of the work, there came a moment when most of the soundtrack was composed. Based on that, the dancer proposed the final choreography to be included in the live performance. In the course of this collaboration, both intervenients decided to also create an audio-visual version of the piece. Therefore, the process used for the creation of the final video of *Neshamah* was divided into three phases:

1. Creation and realisation of the original music and choreography, according to the original purpose of the live presentation, in concert.
2. Video recording of the body movements based on the proposed choreography, to be used as 'raw' footage material in the audio-visual version. This part was done exclusively by the choreographer, which decided the images to be used, as well as the type of movements she wanted to emphasise, either through close/detailed angles or full body perspectives.
3. Manipulation, transformation and compositing of the 'raw' recorded footage. This part was done exclusively by me, as I established a specific type of dialogue between music and image.

The concept of Neshamah relates to the idea of breath as a creation metaphor. Several interpretations of this metaphor include breathing as a source of life, the blowing of the divine breath into the dust of the earth, leading to the creation of life, the rising of the inert body into a new living creature. Both music and dance used such metaphor as the basis for the composition. During the composition process, many of these ideas were exchanged between me and the choreographer/performer and used as the basis for the construction of the body movements and the music associated. Visually, the proposal created by the choreographer progressed organically, from an initial close perspective of subtle movements in the body (or parts of the body), presented with great detail (slow breathing movements, ribs and abdomen contracting), towards a clear view of the full body, accompanied by an increasing movement activity, culminating towards the end of the piece. The evolution from the molecular detail (a metaphor of the dust), and its subsequent path towards the full body construction and life beginning, was thus accomplished.

Throughout the work process for the creation of the piece, a structural exploration was made by appropriating the qualities of the electroacoustic music, specifically the idea of granulation and fragmentation, leading to an investigation of the qualities of the body movements and subsequently the application of several transformations in the manipulations applied to the raw footage.

Formally, the musical idea (proposed by me) of the division of the piece into several sections, where each section represents a different sound interpretation of the breathing process, was confronted by the choreographic proposal, embedded with a more organic growth and teleological character, thus establishing a conceptual counterpoint. This was extremely fruitful for the elaboration of the final audio-visual version. Ideas and concepts such as movement, energy, gesture and texture were frequently part of the exchange and dialogue between me and the choreographer, and although they were many times interpreted in different ways by the two intervenients, they became the basis for the construction of the music, the movements of the live choreography and the audio-visual version. Sound textures may combine as counterpoint to a gestural image, or vice versa; or they may collaborate together in a specific gesture or texture. Many moments in the piece used a complete energetic synchronism between gestures in images and sounds.

In the video version of Neshamah the human body is always present but not always recognisable. The moments where the body (or body details) is clearly recognisable interacts compositionally with the moments where they become obscured or hidden by the effects used in the transformation of the raw images. The musical ideas follow a similar perspective: there are moments when the breathing sound is clearly recognisable, as opposed to other moments when it becomes a subtle suggestion, or even lies hidden in the musical texture. This mutual interest in detail and subtleness, in contrast to more general views and clear movements, was an important basis for the collaborative work.

As mentioned before, phase 3 in the composition of *Neshamah* was the construction of the audio-visual version. It went through a process of analysis of the raw recorded image sequences, choosing the ones that would be most suitable to interact with the music and the structure of the piece and subsequently applying the visual effects and compositing. The transformations applied to the raw material tried to create a link between the musical gestures and textures and the images. The piece tried to integrate the two formal perspectives mentioned before—organic/teleological as proposed by the choreography, and sectional, as proposed by the music.

For those interested, this work is analysed with more detail in the article 'Sound, image and dance interaction in Neshamah' (Oliveira 2017).

Hiromi Ishii

Hiromi studied composition in Tokyo. Electroacoustic music at the Aufbaustudium of Musikhochschule Dresden with Wilfried Jentzsch. Having passed Konzert Examen she further studied at City, University London with supervision by Simon Emmerson and Denis Smalley where she completed her PhD degree. Her research, 'composing electroacoustic music relating to Japanese traditional music', was supported by an ORS Award Scheme scholarship of the UK. Her works have been invited and presented at music festivals including CYNETART Festival Dresden, the Electroacoustic Music Festival Florida (granted by Japan Foundation), MusicAcoustica-Beijing, EuCuE Canada, EMUfest Rome, Música Viva Lisbon, Gaudeamus Netherlands, the International Concert at Musiques & Recherches, International John Cage Festival Halberstadt, SoundTrack_Cologne8.0, Punto y Raya, NYCEMF, TIES Toronto and broadcast by the WDR, MDR, Radio Berlin and more. She was Composer in Residence at ZKM Karlsruhe on three occasions.

Dave: *What inspired you to begin creating audio-visual works? Who or what are your main influences and why?*

Hiromi: I have always been interested in audiovisual creation since I started to compose music, but the direct chance was the Visual Music Marathon in Boston 2007 (organised by Denis Miller) where I watched many works together with my husband Wilfried Jentzsch (also a media-artist and composer). There I obtained various technical information about motion-graphics, which encouraged me to start visual creation by myself. I learned drawing and oil painting a little at high school, but I chose composition to study after all as I was interested in 'effect by sound and music' in films. Having finished a graduate course, I taught at a technical school for sound and recording technique where I composed for exhibitions at aquariums and science museums using synthesisers. Then I decided to study electroacoustic music from the beginning and came to Germany (in 1998). Just after my PhD (in 2006) I began to create the visual part by myself

How do you manage the relationships between sound and image? Are they equally important, does one lead to the other and/or does this change with each project?

In my works, moving images and music are equally important. As visual music, an image needs to develop to be moving images, and a sound needs to develop to be music (also non-Western style music). In my works, both media have their temporal axes. To start a composition I often get inspiration for sound which is accompanied by image, or image accompanied by sound (not a kind of synaesthesia). Inspiration for an image develops soon, but at a stage that needs a temporal structure or an exact timing, the visual waits for musical development, as if the visual wants to 'discuss' with the music. The events of both media have their own character/size/speed (and more factors). Therefore, I ascertain deliberately what element should be given priority. This ascertainment is required at any time during a creating-process, and the relationship between the two media (which leads the other, they confront each other, how they react, and so on) changes at any moment.

How do you approach a new composition—with a concept, a technology, a call/ commission . . .? As things progress, how much do you improvise with the developing artwork and how much do you stick to the original concept/ compositional intent?

I get inspiration for new work often just doing everyday routines such as walking, seeing landscapes and trees, or doing household matters (please refer to the program notes of *Avian* (Ishii 2018), *Kiryu-sho* (Ishii 2019) and *Ice* (Ishii 2020)). Also, I get inspiration for piece B during composing piece A, or even during concert presentations of my pieces. Calls for work cannot be responded to, as I work very slow. My composition method is a kind of 'vector composition'. As my plan is not precisely decided in detail at the beginning, I modify it during its progress, whenever I find better ideas for my concept. A lot of video and audio files are produced and abandoned, however some are rescued and kept in the folder named 'unused' if they appear to be good material for another composition.

I prefer to use taken photos (not to generalise from nothing) for visual material to create abstract images, as I find colour and light/shadow nuances existing in nature are more complex and it is the same way that I use recorded sounds for sound processing (not synthesiser sounds). Also, using concrete materials gives me ideas for the concept.

What are the main technologies that you use for sound and image and/or do they change depending on the project?

- Visual: Studio Artist, After Effects and Trapcode.
- Sound: SoundHack, Max MSP (Own programs), Audiosculpt, Nuendo.
- Other programs can be applied depending on projects.

Can you say more about the approach you took in creating Aquatic, 2016 both artistically and technically? What inspired this piece?

When I started to use Trapcode I watched many wonderful 'demo' videos of it, but they were accompanied by music that had nothing to do with the visual part. This motivated me to create a visual music piece that has a close relationship between visual and music.

After several tests, I used a photo of fishes, which I once used for my older piece, *Refraction* (Ishii 2013). The expression of colour, light and shadow in this picture transformed, appeared most delicate and complex among other materials tested. Also, the transformed images reminded me of underwater living such as jellyfish, mantas and others. At this moment the concept for this work became clear.

On the other hand, I had an idea for an acousmatic composition with underwater sounds and those in the air. I thought these two projects would fit together.

Fortunately, I was invited as Artist in Residence at ZKM Germany a couple of months later. There I had a chance to complete the music part of this piece using a Zirkonium 3D sound system with 44 loudspeakers.

In this piece, the visual speed of transformations and movements is equally important to the musical tempo. There are two independent time axes (of visual and music) and they are applied to structure and dramaturgy. A narrow frequency range of the beginning sound leads to a mid-low sound, then it reaches a super-low sound, whereas a small figure in the visual part develops larger and is more layered to cover the whole screen. At this point the size of visual events shifts from 'figure and ground' to 'multi-layered planes', and there is 'no background'. At this stage also the music uses the full range of frequencies, however, true acoustic shift in music is at the moment of the 'big wave' sound which is not heard underwater, but in the air.

Tim Howle

(In-depth interview)
Tim is a freelance composer and Emeritus Professor of Contemporary Music in the Department of Music & Audio at the University of Kent. His first post was

at University College Salford (1990–92) where he lectured in composition, 20th-century music and sound recording. In 1992 he moved to Oxford Brookes University where he became Director of the Electronic Music Studios. From 2000–10 he taught at the University of Hull in Creative Music and took up his post at Kent in 2010. His academic career focused on composing electroacoustic and acoustic music, and the relationship that these areas have with other arts subjects. Tim continues to compose primarily sonic art, acoustic music and music for experimental video with Nick Cope.

Dave: *Can you tell me a bit about your background and how you began composing for audio-visuals, rather than just music?*

Tim Howle: I was always aware of it. Composing with recorded sound was only half of the picture. But of course, the picture element, the visual side of it, was always missing in the way in which sonic art was described. 'Cinema' and 'cinema for the ear' are obviously words that are used very often, but the visual aspect of cinema was rarely discussed. It was always through omission that the visual side was there, so the point seemed to be that you had a lot of freedom working with sound because the visual side was missing. Understanding where Schaeffer comes in is actually very important, but it's surprisingly straightforward once you get your head around it.

The other thing that got me into this was collaborating with people. I think a lot of people, particularly going back to being a postgraduate composer, see themselves being filtered into this area where techniques and the need to be appearing to be good at what you do, means that you see yourself as a lone practitioner, stuck in a studio with no daylight, on your own—being great at things. I didn't realise, until I started collaborating with people, a lot of the collaboration gets things moving. I've collaborated with Diana Bell, a visual artist in Oxford, who presented a set of questions to me, for example: one of the things she was interested in was the 'presence of absence'. So she did a piece where there was a suit of clothing that she'd left out on a lawn during the winter to make it look weathered and decomposed. She had lost her father. She said she was going to hang the suit on a wire frame, it looks like it was being worn, but of course, there was no body in it. That's the point. My job was to come along with music that wasn't present and, although I find it a bit contrived, I got the idea, so it gave me a job to do rather than just doing what I wanted to. Then I recognised that I was actually collaborating with people more often than I thought.

At Oxford Brooks there was a guy called Graham Ellard and he was another fine artist, his thing was video and again that was fun, but I was very much driven by his agenda again. As a Fine Artist he had a lot to say about the philosophy behind what he was doing, whereas musicians very often haven't. At times, I just wanted to make music and found the idea that I had to talk about it constantly, slightly irritating. So I did a couple of things with him. One was based on a video shot through a pane of glass, as if you were looking into a space. Once it's been shot there were a few actors employed to move things around in the space or sit on chairs and come in and out of doors. And then he edited it afterwards; it becomes a pseudo document. Then I produced sound, which was both 'in' and 'out' of the space, and then the fun part was when we projected the piece onto the window from which it was shot, but from the inside, so it looks as if you're looking into the space again, but you're not. You're looking into a recording of the space. It was at the Royal Festival Hall and walking through the building you walk past the window, you look in at the impossible things that are happening. So again, from an acousmatic point of view, I thought that was good fun. The collaboration ran out—I went up to Hull University where I met this guy called Nick Cope . . .

I remember this from a while back. He was doing some visuals in one of the local cinemas?

Yes, I didn't know that much about him. He taught digital media, video, that kind of thing. And I sat in the cinema and saw what he was doing. He was mixing two streams of information live—using an analogue video mixer. Just basically two levers. Moving them backwards and forwards, and he was cutting between images of travelling, literal images, things shot from the front of trains and things shot from helicopter windows as they went past tall buildings and just simple Winamp animations. It generated feelings as if moving forwards all the time and it seemed to me that was acousmatic, but a video version, a reversal of my role. Afterwards, I suggested that all of his language dealing with video editing and my language dealing with audio editing, were just converse versions of the same thing, so that's it, we were off.

Yes, and that's ongoing to this day, isn't it?

Yes, it's become more difficult because a lot of the original material that Nick had was basically scratch video things from the 1980s, so the fact that they were shot on Super 8 didn't matter. We had this odd situation where we were going to SEAMUS conferences in the States where everything was glistening, high-res rendered ani-mation material and our stuff might have a hair in the filter and bits of dust in the frame, but of course it's all part of the what we do. It feels European, almost like a razor blade approach and that's continued. What we need to do really is generate more material, and if anything, pieces that we've been working on backwards and forwards for a while, are much more high-res. A possible new piece is based on a piano, so Nick filmed a Disklavier, up close, very high-res, pristine images, and I've recorded the piano—it's slowed us down because the fun, the dirtiness of it is missing.

Yes, it's better with that immediacy rather than thinking about the process a bit more. Is this the latest piece that you're working on?

Yes, but it has been going on for a while. In a way it asks questions about how people collaborate. Because we don't mush along very well—in that we discuss every event as if we needed to agree on how that event would unfold. I feel quite comfortable just working to finished films in a way, we agree on what the film is about and how the sections might work. But actually Nick puts it together. We have worked the other way around. I presented Nick with finished music for *Son et Lumieres* (Cope and Howle 2006). It is a piece like that, and it also shows up the primacy of the video side of things really well. If that works, it's quite easy for a composer to move in and out of the frame or stop/start, or keep things synchronised or non-synchronised. Whereas to work the other way around doesn't seem right, and maybe that fits with the way we perceive things. Visual elements are immediate and audio things less so.

Is it always video/visuals first? From what you're saying it's not and it sounds like one way is a bit more natural than the other. It makes more sense to start with a video.

Yes, and then he can always edit. If I do something, let's say a section needs to be longer—that's fine, we can do that. The other thing is, I think a lot is serendipitous. To discuss it too much kills it a little bit. Materials might or might not fit together. Moments can be capitalised on through editing.

That covers a lot of different approaches. There can be video first, then you compose the music. There's music first, then you create the video, or it's on a more granular level where you gradually share ideas and build it up and that's the one that's more time-consuming and complicated maybe?

It is and I suppose the other one would be . . . I play with software where what I'm actually doing is just moving things around on the screen. The video and sound that's generated fit together on a gestural level immediately. It's too easy. I go to conferences and I think I'm looking at something that's very impressive, if it's an algorithm delivering data, there can be beauty and I would never take anything away from a composer or visual artist who generate their own software or indeed uses Max or Jitter, for example, but there's something in actually looking at the material, making decisions and although it takes longer, it's just a different way of looking at things.

Yes, I think that is another way and there are two approaches. One is the composer doing everything, making the decisions, 'composer rendition' is maybe one way of describing it, and the other is the algorithmic side and there are a lot of apps out there now online where you can put something in and it'll generate and it's quite good.

Yes, and why not? If we want to write for a violin, we don't go into a forest and cut a tree down and make a violin, we do use software. It occurred to me while having some fun using logic to put some ideas together. I was playing around with one of the drummers in Logic where I can call up using a simple AI drummer, and I can almost speak to the drummer in the same way that I would have 40 years ago to a drummer in a band. I might be saying to the drummer 'could this bit be simpler and louder and could the other bit be more complex and lighter'? And now I can do that and I don't feel that's cheating. I feel as if I am almost talking to a drummer; alternatively, I'd have to physically go and find a drummer and ask whether they would collaborate with me. Why do I need to know how to programme things if I can basically just talk to Siri and it just does it for me? I'm still in control of the creative process. (I'm stretching things here—I do like to know what is going on under the hood—as far as possible.)

Yes, perhaps there's an aspect of uniqueness about what comes out of it. A lot of these apps will produce generic-type materials, won't they? I think the more control you have over the algorithm, the more you can tweak it, I suppose, the more the composer input there is.

To me it's part of just getting older. When I started off on a Revox B77, it was all done by hand, to the point that if you wanted to start three machines at the same time you'd have to actually wire their start buttons together by getting a soldering iron and doing it yourself. On the other hand, I saw a Native Instruments app that was basically choir sounds on an XY pad: one corner sounded like Lutosławski, another Ligeti and so on; moving the curser would morph between these very sophisticated sound worlds created by somebody else at Native Instruments who knew how to write this material. From a gestural point I could just sit there with an iPad and just react to the visuals doing this kind of thing, but you know, from an education point of view, we should address who needs to be able to write for choirs so that they can put their own material into those apps. Or do we all just do the homogenised thing? And let's say I did it and I spent a year writing for all these choirs and recording and then bringing all those recordings into the studio and then morphing between them using transformation techniques and if I did that, would it be any better than just finding the Native Instruments app? (Hopefully it would!) I think that these tools produce interesting results without much effort—not always a good thing. Going back to Nick, doing things ourselves, genuinely using microphones and using cameras, it feels worthy. It feels like there's a point to it because it is ours. Recording and editing is 90 per cent of what we do.

I think there's always an element of performance in composition. OK, you're performing on a laptop. Maybe it's different to playing a guitar or other instrument . . . but that's the unique side of what you're doing. You're performing new ideas, new material, and you're capturing that as a composition.

Yes, I agree completely. I think that you're making creative decisions based on your own sensibilities.

Going back to your collaboration with Nick, you also described your compositions as electroacoustic movies, is that still a name that applies to your work?

The idea is that the language that exists audibly when composing used to be called electroacoustic music. Our term is sort of an in-joke. And the joke is that it can't possibly be electroacoustic music. If video is added you're spoiling all the fun in one sense, and introducing a whole new raft of different sorts of fun. The word 'movies' was adopted. The word 'movies' suggests that we've gone to the movies. It's a light-hearted thing, it was supposed to be a witticism. The serious point was the 'audible silence' and a 'dark screen', where the screen would contain no information on it at all, and a counterpoint is formed. They balance each other, starting off from that point of view; looking at silent movies the sound is better because the way we conjure up images within our own minds is in a more sophisticated way than being presented with concrete material. We synthesise things that would be impossible to produce, and likewise we might synthesise visual images in our minds, if we were just listening to sound and not having visuals to look at. So somewhere in between the two, the idea that there's an electroacoustic approach to video is still there.

Do you still see some reluctance by people from the electroacoustic community in accepting video/visuals added to their electroacoustic compositions?

I know people, ha ha. The language itself is so nuanced, and the parameters that are used when using acousmatic principles are so well talked about. They are talked about to a point where it is quite repetitive. By about 1990 it was repetitive, although it might not be new to somebody who's been doing it for a while, it's new to the next generation, it reinvents itself and we seem to be describing more finely graded arguments, but they're basically the same arguments. I've always been about broad brush strokes. I think the basic ideas are in place and there's a natural protectionism regarding these ideas. An electroacoustic composer will see the problems that are going to arise when introducing visuals. Putting two opposites together might just then cancel each other out. Having said that, most people agree, in terms of first principles, that the act of putting those two artforms together is very exciting, something is happening. Something that triggers all sorts of reactions, there's something there. Also, to some extent writings about the audiovisual also plateaued quite early, I think, and it's difficult to talk about now without going into details that are more to do with each individual work, the basic groundwork has been done. I don't see anything wrong with that. In order to compose for a string quartet, I wouldn't have to continue to describe what a cello is and what a viola is, or what an interval is – it's a known, we don't need to keep going there.

Do you do you still just make music or are you purely an audio-visual composer?

There's a chap called Jos Zwaanenburg from Holland. I don't write enough for him. When we do get together it's much more of a collaboration than he admits. He's a fantastic flautist. He runs the live electronics post-grad department at Amsterdam Conservatoire.

How do these collaborations begin?

Nick was a chance thing, in the sense that he got a job at the same institution I was at. When I was an undergraduate in about 1985, Jos was a similar age to me.

He turned up doing a tour of what was then described as live electronics and flute. He was saying things like, here's a flute note, here's a flute note played with multiphonics. Here's that multiphonic being taken through this processing to expand it further. That was the language used at the time and I was quite ambitious. So I wrote to him and it just triggered something. We've only done, I think, four pieces over the years, but they work for us and they give us a platform that lasts quite a long time. We get lots of performances.

I think a couple of things about that. One is that if you're going to write for string quartet that's OK because there are lots of string quartets around. But if you write for string quartet plus an accordion, you're never going to get a performance. You might get one, but it would be hard work. There is that path of least resistance thing going on. And I don't think I've ever been cynical about collaborating with people. It's always been serendipitous, I've bumped into people and realise that I find what they do very interesting. So I'm quite comfortable with that, but I'm aware of the fact that from a purely practical point of view, a good collaboration works in a sense, because both parties are pushing the work. So Nick, for example, pushes the work in film festivals as I push it in music festivals. The great thing about Jos is that he is completely self-contained from a performance point of view.

What's the process you go through when adding music to visuals? Are you looking at qualities within the visuals or are you using them as a general kind of influence or inspiration? How do you manage those relationships between sound and image?

I suppose I try and separate things out such that the quality of the visuals suggests things anyway, so gestures or hues or graininess in the image or linear images, the things that gradually unfold are capitalised on.

Do you feel you're sort of interpreting these visual aspects in a musical sense?

Yes, it could almost be a score. It's almost like a time-based version of a Kandinsky image. Colour I don't think of in a synesthetic way, but I'm aware that it is there. To me C major is, for sure, light blue. That is going on clearly and then the acousmatic thinking kicks in, so some things are in the frame and some not. Out of the frame the music almost becomes incidental, so it becomes free and manipulation becomes freer. Material in the frame can then be reworked so it loses any attachment to the other objects around—it then might leave that environment. Ambiguity is one of the things that makes music: where does the object sit? What does it belong to? Proximity is akin to consonance/dissonance. Without this there is very little composing going on. Sometimes relationships are obvious, sometimes they are hidden, sometimes it's difficult to tell either way and I think that's the thing that draws somebody in.

And I find sometimes if the video is quite busy, the music can be quite minimal. Maybe if you're literally interpreting every gesture in the image, it can become overwhelming and sometimes it's the opposite works.

I agree completely—if suddenly we see an image which becomes quite busy, as if there's some energy in it, but the sound comes a little bit later. It feels quite natural. Sound might not travel as quickly anyway; it enters your mind half a second later. It can also be because one object excites the other.

There are so many different ways you can work with materials. I don't know if it's more because you're working with visuals but it gives you different ways of interpreting it than it would be from a purely musical perspective. How much do things change as you go along? Do you stick to an original concept or idea or do they gradually evolve as you go through?

The way I work, I tend to spend a lot of time looking and listening and preparing, analysing, so by the time I'm actually putting it together, I'm two-thirds of the way there already, I know what I'm trying to do. I think that one of the things that happens with this kind of collaboration is that there is an inertia because the amount of effort that it takes to actually say to both parties, 'let's change this bit' . . . sometimes you think, well, can we work around it? I always thought that composers need to just be pragmatic sometimes. The art is everything and maybe you've got to spend years on a piece until it's absolutely perfect . . . but deadlines are good too, if I've got an extra month I would make a particular section slightly different. But sometimes it just needs to be composed. A theme might be that composition is also craft—techniques develop through repetition. Thinking musically doesn't have to be extra-musical too.

Has the technology changed, have your working methods been changed by the software that you're using?

I always felt that categorising sounds was a good thing. Because clearly what needs to happen at some point is that, if several sounds are used, they either need to be related to each other or not, in the same way that notes in a melody are 'close' to each other or not. I can be quite systematic, and if I have lots of days where I just need to do some work and don't feel particularly inspired, I'm just quite happy working on sounds. I've got sounds called 'filtered-clatter' in a folder called 'clatter'. If I work on clatter sounds and nothing much happens then I move to a different folder. With *Open Circuits* (Cope and Howle 2003), Nick and I had been working on prep for a conference anyway, just on the organisational side of it, so I realised I've got three or four days left, and then I'd only got the evenings to compose in. I was working at home in a room full of laundry with two speakers poking out between piles of T-shirts and trousers, just grabbing things from the folders of sounds quickly and just syncing things up. And I'm thinking OK, that worked, so I'll take that phrase out, use it, but also copy it and move it to some area where similar images came back and then reworked it. There was a sense of development. Regarding background elements, I did the same thing but on a slower rate and then categorised those objects into folders. Then I mixed sounds down to more complex objects—repeat . . . repeat . . . The piece has a coda. It's almost as if two-thirds of the piece happens. About two-thirds of the way through it seems to speed up, but also become lightweight as if the material has less energy, almost as if it becomes an afterthought. I thought, well, I've introduced new material near the end, that's a good thing. I introduced something melodic that hadn't been in the piece before, signalling this change then filtering gradually all low frequencies out until that material formed a whisper.

It's very quick moving, the sound with rapid cuts and edits, and then like you say two-thirds of the way through it becomes more droney and pitch-based and ambient. Maybe another way of describing it?

The decision for the pitch-based sounds was, two minutes, go to the right folder . . . The pitch-based sounds become incidental music and successive lower octaves are introduced to give it more oomph, and then the material in the frame gets filtered the other way, so they gradually diverge and fade out, it's clear that you're coming to the end of the piece.

It works really well, and I think of it in terms of gesture and texture as well, like the first two-thirds is very gestural, everything, it's very synchronised. What happens in the visuals is reflected quite strongly in the music, but after that it becomes detached in a way; you're not responding to everything that happens visually or you're responding to it in a more textural way.

I learned a lot. I learned that being preoccupied with how things work doesn't always propel a piece in the right direction. It was just a simple idea that worked. When I talked to Jos about composition, (he teaches composition as well as doing live electronics) he might say to students something like 'this material is A and that's B and make A become B for over a period and don't overthink it'. To actually be able to do that at all requires real technique. It has to work—something out of the trajectory will move the piece on. Notions of 'A' and 'B' might get left behind.

Another way of working with audio materials is a bottom-up approach, where you've got these folders and sounds and you start arranging them and see what comes out. Another is top down where you've got an idea or concept for the piece, and I think when you're working with visuals it feels a bit more top down, but at the same time you're doing it the other way round, sort of meeting in the middle?

That's a good way to put it. There's a lot of tension there—in trying to get into the middle.

Your main platform is Logic then?

Yes, I never really got into Pro Tools. It didn't seem to do anything that I couldn't do elsewhere. Logic does seem a bit straightforward nowadays. Some think it's GarageBand 'Plus'. It's very easy to use. If I'm using MIDI at all to trigger things, then I use Logic for that. If I was writing something for flute and other sounds, I tend to use Logic again for the same reason. I can move things around instrumentally quite easily, but that's about it. And then I use Adobe Audition a lot because it feels like a tape studio. Zooming into a single frame or sample is easy. Lining things up too, it just feels precise and controllable, whereas Logic feels like I'm moving around bags of flour. I'm looking at the new spatial possibilities in Logic— could be fun. As ever, I'm very much an end user. I might have apologised to Todd Winkler once at a conference in that I came out and said that I did not do any coding and so on—he said something like: 'it's OK you're a composer'.

Back in the day I used to use Composer's Desktop Project (CDP), which is a text-based command line suite of programmes and in fact I've just got an old PC that I can use and actually do some of those transformations again. It does do things that are very difficult to do with plugins. There's something in the CDP system called 'Sausage' and I am looking forward to that again too. I do use Audiomulch because in my non-Max-MSP world I can specify complex things very simply. I can create structures and control things through automation very easily with breakpoints almost like the CDP system. It's something like a version of Max for people like me. One of the things that it does really well is that it's got something called a metasurface. It saves scenes, drops the scene onto the surface, sets something else up that you want to transform to, saves that as a scene too, drops that somewhere else on the surface then just using your mouse you can morph between the two or more points There might be 50 different parameters between the two scenes that you're controlling. Gesturally it is very interesting.

Jon Weinel

Jon is a sound and image artist with a wealth of creative outputs who has also written two books (2018; 2021) discussing his work. He has presented his music and audio-visual productions at international concerts and conferences including ICMC and ACM Audio Mostly. He is currently Senior Lecturer at the University of Greenwich, London, UK.

Dave: *What inspired you to begin creating audio-visual works? Who or what are your main influences and why?*

Jon Weinel: I think it came from a mixture of things. First, growing up with computers like the Atari ST, from a very young age I became interested in making graphics and sounds on the computer. That led me to become interested in making electronic music on the PC, and studying music technology at Keele University to further that interest.

At Keele, audio-visual composition was one of the areas being taught by Diego Garro and Rajmil Fischman, so I got exposed to more of it there and had the space to start making my own compositions. As a result I was aware of thinking related to spectromorphology and considerations for mapping those morphologies into visuals. However at the same time, I had become very interested in psychedelic music and art, so it was obvious to me to combine electronic music making with psychedelic visuals. A lot of my undergraduate audio-visual work was a mixture of video game references, psychedelic visuals and electronic music, and I continued those into my postgraduate studies and PhD work. From there it continued to grow and I became more interested in real-time work such as VJing, as well as working with game engines and VR.

In terms of specific influences, it is a really broad range of stuff, but to pick a few: Jeff Minter's light synths, Harry Smith's visual music films, rave visuals video tapes (e.g. X-Mix series) and psychedelic video games like LSD Dream Emulator. Musically, genres like old skool rave, hardcore techno, psychedelic rock, electroacoustic music, surf rock, dub reggae. Demoscene work is also a significant influence on my music and visuals.

Which one of your compositions or performances would you recommend to someone new to your work and why?

Probably my most recent work *Cyberdream* (Weinel 2019), which is a VR rave music experience for Oculus Quest. I consider this my best work to date. The piece is very synaesthetic in the way it combines rave music, visuals and interactive audiovisual sound toys. Alternatively, a recording of one of my Soundcat VJ mixes or possibly the fixed media work *Mescal Animations* (Weinel 2013).

Did you start as a single discipline artist, for example, musician/video artist and how has that coloured your experience working with multiple media? If you always worked with multiple media, how did you simultaneously develop skills and interests in both media?

I have always worked across multiple types of media, I always find it more interesting. In terms of academic studies, at Keele University I studied music technology & visual arts as a dual honours programme. Later I did an open degree (where you can choose modules from any field) with the Open University with modules in psychology, mathematics and computer science. I work with code, sound, paint and graphics and often combine aspects of these in my work. *Cyberdream* is an independently produced VR work, where I created the sounds/music, programmed it and also made some graphical elements by digitising hand-painted artwork.

How do you manage the relationships between sound and image? Are they equally important, does one lead the other and/or does this change with each project?

I think usually there is an interplay between the two, both within projects and between projects. One project might start with visuals and that inspires some kind of sound/music or vice versa. The feeling and excitement of music is often the starting point for a lot of my work, and that leads to the visuals. But within that, I might actually make the visuals first and then create sounds/music to go with it afterwards. So the visuals in a sense flow from the music, but both are intrinsically related.

What is your approach to live audio-visual performance and how does it differ to composition? For your live pieces, why do you choose to perform them rather than compose fixed media? What are the main challenges during performance?

Not all of my audio-visual work fits into the 'live' category, as many works I produced were fixed media, and some were software-based projects that might be experienced by individual users in an exhibition or as an installation. My main live audio-visual work in recent years would be my Soundcat DJ/VJ performances. These are essentially DJ sets of existing music, combined with original visuals. I fundamentally approach this as a DJ set with visuals, as this project was in many ways going back to my roots in DJing hardcore rave music. DJ sets are usually about creating some kind of journey through sound, and a lot of it is about the energy and excitement that comes from the rhythms and the bass. Visuals add a complementary psychedelic component. The live experience is more fun, I think it's more enjoyable for the audience in many cases.

The main challenge for audio-visual live sets like this is the extensive amount of preparation work that goes into creating original visuals to fill a whole set, and finding stable software/hardware configurations that won't have any drop-outs or latency issues.

What are the main tools and technologies that you use for sound and image and/or do they change dependent on project?

They change, however there are some constants. On the sound/music side I am a Renoise user, the tracker interface clicked with me when I tried it many years ago and I find it much more enjoyable as a way to make music on the computer. In terms of visuals, I use/have used a range of software such as VDMX, Processing and Unity, as well as analogue equipment such as 8 mm projects, acrylic paint and an airbrush. I'm currently doing some work with Unreal 5 and so I may end up using that for something in the future.

How did you go about developing the materials for your VJ performances? What links or influences were there between the sound and image media? How do you manage/explore these relationships during performance? Describe one specific VJ performance you undertook.

The visual materials for the VJ performances were developed mainly when I was living in Denmark doing a postdoc in Aalborg. I had a nice, bright, empty flat by the fjord, and had a bit of spare time to work on my projects. I got into making visual loops in Processing and mixing them with other types of material I made with direct animation, or quite crudely produced stop-motion animation. I was really into VDMX at the time, it is a really nice piece of software that is efficient and really reliable while being modular and really customisable. I would make visuals through them into VDMX and spend hours watching them/playing with them while playing lots of different music. Through that, you get to see how different combinations look/feel together. Visuals can work with many different kinds of music, but often there is a link between the amount of kinetic motion in the visual and rhythm/ percussion. Rhythmic music requires visuals with lots of movement essentially. By exploring combinations you find what you like, what you want to see/hear and find most satisfying. Eventually you can shape that into a performance, and hopefully other people like it too. I got some good feedback so I think it works okay.

What are your future plans for composition and/or performances?

At some point I may do a follow-on work from *Cyberdream* exploring related ideas. I don't plan ahead strongly and creative work normally happens when there is a strong urge as a synthesis of current interests and experiments, which makes it hard to predict what I might do next.

Bret Battey

(In-depth interview)
Bret is an audiovisual artist and Professor of Audiovisual Composition in the Music, Technology and Innovation Institute of Sonic Creativity at De Montfort University, Leicester, UK. Bret has an impressive catalogue of audiovisual productions and academic research from work spanning more than two decades. He regularly screens his compositions and presents talks on his work at international concerts and conferences.

Dave: *How did you get started in audio visual composition? What were you doing before that? What were your main interests?*

Bret Battey: So it starts with my undergraduate degree in Electronic and Computer Music from Oberlin Conservatory. The program's name was Technology in Music and Related Arts. So it already had an ethos of connecting with other artforms. But while I was there as an undergraduate, I fed myself through graphic design. I worked for the graphics department at Oberlin, in helping in their transition to desktop publishing and doing layout and stripping things up for the printers and so on. And after I completed that degree I went back to my hometown of Seattle and made a living by temping.

Were you a music composer at this point, would you say?

So yes, with that degree from Oberlin, we had focused on electronic music. So I had that formal training, but I had less formal training on graphics, but that's what I was actually feeding myself with to a large degree, and even became a Webmaster back when they still used that term for the overarching phrase. So by a certain point I started to think about going back to academia and pursuing a masters. And I was really imagining doing something that included visuals at that time. I was thinking more in terms of live performance, with audio or visual light processing sculptures. Say it could be warping video or just raw light and casting that up, like dramatically electromechanically controlled through the same algorithmic systems I'd be using with the music. That was the vision I had in my mind at the time. And so then I stepped into my masters at the University of Washington in Seattle. One aspect of that started looking at computer graphics. At that time they had Silicon Graphics machines that we were using for the computer music and computer animation so both my masters thesis and my ultimate doctoral thesis from there were audiovisual pieces. But as a masters student, I was thinking this is kind of foolish.

I mean Computer Music, music technology itself is so deep and a lot to master, you know, so what am I doing, going and making visuals? I also was thinking I should collaborate with video artists, but the video artists I talked to, I couldn't find any that understood where I was coming from as a composer, that I was interested in abstract texture and gesture. And some video artists tend to be very conceptually trained now. In Seattle at that time we had people, but they weren't interested in the really deeply crafted moment-to-moment kind of thing. So I said, OK, I'll have to do one myself so that I can demonstrate to a video artist what it is I'm trying to do. But then I found it so rewarding to be working in both mediums together that they feed each other creatively, I said, well actually this looks like my path. So that piece was *On the Presence of Water* (Battey 1997). That would be my first one.

So it's the convergence of technologies? You were using the same platform to make music?

Let's see, at that time, actually, the music didn't use a digital audio workstation, so the music would have been CSound controlled algorithmically with Lisp. And then I think at that time, we were using something out of Princeton called RT, which ran on Linux. You could basically list all the sounds that you wanted to have in your mix, what time they start, what their amplitude is. So it's kind of a text-based mixer, if you will, and I think that's what I developed that mix in. I was familiar with things like Performer and MIDI at that time. But the video was in Premiere, I believe, a very early version and in this case, I was taking video footage and manipulating it. Those kinds of materials and still images, really zooming in closely and moving across still images.

It's like the musique concrète version of the visuals, having video as visual objects, and transforming them.

Exactly. So the first two pieces I did were really more in that mindset. Then after that, a different approach where the video source material is there, but much more hidden.

So you moved to the UK at some point. Did that instigate the switch for some reason, or was it completely unrelated to the move?

I finished my PhD at the University of Washington. Then I had a Fulbright Fellowship to India for a year where I was studying Khyal (a North Indian classical vocal music tradition), taking lessons in that and creating software for modelling the pitch curves and ornaments in that tradition. But I also took my first intensive Buddhist meditation retreat, 10-day silent retreat at the end of that stay there, and the combination of that experience really threw me as an artist. When I came back, I felt very disoriented as an artist and my relationship to the Avant Garde, to Computer Music and so on. And it really took me two years, I think, to rework myself, and in fact a piece that does that is *Autarkeia Aggregatum* (Battey 2005). That was the arrival of a new artistic signature, and in many ways both technically, but also the inner space that was coming out. Part of what happened there on the visual side is that I was teaching a class on generative techniques, visual art as a postdoc at the University of Washington, and I made an error while doing one of the code examples. I was supposed to show how rotations worked within the Processing language. And I just made an error in that and instead of rotating a box made of dots around, it spewed these dots around because it was actually rotating the 3D space before drawing each dot, and that's explained in detail in one of my articles (Battey 2016).

There is feedback where you get trails of these dots?

Well, it's easy to do in Processing, you know. Just tell it to not erase, or only partially erase, the previous frame before drawing over with a new one. This discovery and this algorithm really fed the work all the way up through *Clonal Colonies* (Battey 2011) in a very direct way. The first piece which does that is *cMatrix 10* (Battey 2004), which was an installation piece and it's actually one of the very few pieces I have where the image came first and then the audio later. But in the midst of doing that piece and taking this algorithm that spins these things in a 3D space which then collapses back to 2D, I was also seeing the importance of slightly randomising the points. So in *Autarkeia Aggregatum*, instead of doing all this complex 3D spinning, it backs up and just says, here is a grid of points, we're drawing from a source image, and then all we're doing is randomising their offset from a centre point, randomising a rotation around their centre point, and that in itself was enough to get some very complex results. So you could either have an image cohere or break it up into a flow of points.

So, it's almost like a particle system-based type approach to visual composition.

It is. It can have some similar things [to particle systems], but it has its own distinct character.

Yes, I think that thread, that idea where you've got very detailed points that come together or disperse, that technique flows through your work.

Absolutely, and I mentioned that in reference to *Autarkeia Aggregatum*, in particular, because it partly came out of the impact of the intensive Buddhist insight meditation practice. Part of what I experienced on that retreat for the first time was the sense of the solidity of the body dissolving into just the sense of the flow of particles. And that's a common impact of intensive insight meditation work. And so it's not that *Autarkeia Aggregatum* is somehow depicting meditative experience, but I think my sensibility shifted at that time in response to the meditation work, and so I was drawn to this particulate sense and the fact that you could dissolve something outward and so on.

It's interesting you say that because one of the ideas for this book is based on an eastern philosophy about an elemental system where everything is created from spirit, air, fire, water and earth. So my chapters are based around those things and the interviews are in the air chapter. It's a good way of conceptualising the book. It feels like there's a little bit of a connection there somehow with what you are doing.

Yes, air and fluidity, you'll see the term fluid coming up in this concern I developed for this idea of a 'fluid audio visual counterpoint' as I call it. But of course I work with my intuition that fluid is the term, but what does that mean actually? It's still something I'm working on and I think the other aspect of *Autarkeia Aggregatum* is it's not based on cuts and edits of material. It's one continuous visual process. From beginning to end with parameters being manipulated and that I found, I created, a very particular kind of character. You know, it's not like you don't get hard cuts. Something is always transforming into something else. Which again is another thing that has resonance with the insight meditation experience.

So that piece was entirely algorithmically created from start to finish, it was one rendition of that algorithm?

Of the visual algorithm, but it would be misleading. It is a visual algorithm, but it's parameters I am controlling entirely manually. So the big leap I made there is that I was originally generating the ideas in Processing and Java. But then I said that I want really fine time control and so, thinking as you do from working in CSound, I'll manually define control envelopes for the parameters and I realised that it was just way too kludgy of a way to move forward, incredibly slow, especially if you want to create something that's going to link tightly with the image. So I took the risk, and I decided at that point to take the time to figure out how to make plugins for Apple Motion. I created a plugin that had this point-spinning algorithm and then I could use Apple Motion's GUI, really refined rubber band tools with Bézier spline curves for time control and everything, to control the parameters of my algorithm. And that was the real breakthrough that allowed me to do this incredibly fine and precise control of that visual process to get it to work with the music.

Can you tell me a little bit more about that plugin development process? I assume that you get a series of parameters in Motion like you would with any other visual filter for example.

Yes, basically I define my own visual filters so I could also use Motion's tools to edit, stretch, zoom in, move around on the video source or image source that I was using and then also very finely manipulate the parameters of my video filter.

And are you still using this technique?

Yes, at least up to *Estuaries 4* (Battey 2021). At this point I'm having to decide, I feel like I need to move in a new direction and it's always been a royal pain along the way because every time Apple makes any significant change in its system or in the Motion software development kit, my plugins would break. And then I have to spend months trying to get things to work again. So this last time around they've gotten rid of OpenGL. *Estuaries 4* took me almost a year to rewrite the plugin to work with Metal, which is incredibly difficult, and it's all done in Objective C.

With the Estuaries series, it sounds like you feel that's coming to a conclusion in some sense?

Yes, in some ways with *Estuaries 4* it feels like there's almost nearly a 20-year line of things, and I'm feeling a need to search in a new direction, although I'm doing one more installation now ['Traces, Molten' for the Sep 2022 British Science Festival in Leicester], so I'm still using the tool I use for *Estuaries 4* and that's explained in more detail in the chapter 'Technique and Audiovisual Counterpoint' in the Estuaries Series (Battey 2020). In my Indian music modelling, part of what I had to learn about was the mathematical domain optimisation. Which means when you have a multivariable problem, you're trying to find the best settings for all the variables to achieve a certain thing. In this case I was trying to get a Bézier spline to fit to a voice pitch curve as best I could. So I learned about this classic method called the Nelder Mead method. Something crawls around on the surface of your mathematical terrain and tries to find a peak or valley to do the optimisation. So I had the intuition this could be interesting; let's just visualise that process and display it. Display all the little polytopes that it creates. The way it works on a three-dimensional problem, it draws a triangle, looks at the values of the three corners of the triangle to try to figure out which direction to go, and then it flops and creates a new triangle, remeasures, decides, flops, creates a new triangle of a different shape and this is how it crawls around on the territory. It's worth thinking of it as an agent. You start somewhere and then it draws triangles to crawl around trying to find a peak or a valley. So normal practice, you would just have one of those to solve the problem, but what I'm doing is just putting up thousands of these agents, and they're all crawling around doing sometimes a pretty bad job of trying to find the brightest points in the source image.

There are resolution issues associated with that. You are asking a lot from the displays, I guess.

Yes, I learned from experience to try to render as high as I can technologically and usually that's going to be a bit ahead of where the display technology is. I'm really sad now, like *Autarkeia Aggregatum*, I really would like to be able to render that at a higher resolution now, but it would take months, maybe even a year, to figure how to recode it and probably wouldn't look the same.

I find that natural dating of pieces is something that you kind of have to live with and they have their own certain character and you get used to it after a while. I find that even with music I feel I want to go back, but really you just have to leave it and accept it for what it is. For the Estuaries compositions, for the visual side of it, is it fed by another image?

Yes, pretty much all of the pieces from *cMatrix 10* are fed by an image.

How do you decide on the original images?

I'll use the term impressionistically: I could say 'I think this looks like this could be a useful colour palette here', or 'this is a useful texture'. If I want something that's fairly noisy and continually changing, I use the surface of water. Or maybe I know I want some really strong colour contrasts. It would be a little bit like a Wizard of Oz behind-the-curtain moment to show the source video to people, although I have done that occasionally. Often you can strongly relate [the source video structure] to the high-level structure of the work and then it's the manipulation of the [effect] parameters that's providing the more detailed level, structural manipulation.

Yes, I think there's a similar process happening in the sound as well. You're using original source material, which is fairly ordinary and then you transform it through an algorithmic process. You start with MIDI sounds and then you use convolution to enhance them?

Yes, in fact, I'm trying to find out a way between two worlds and one is on the music side, the event-based side. It's always easier to get algorithms to work that way, [generating] notes, note events. And then, of course, [I am] deeply steeped in that aspect of the electroacoustic music domain that is not event-based, that's gesture- and continuum-based. And so I've always found myself working between these. So this technique, the variable coupled map networks, lends itself to this event-based approach and I am just trying to find various ways to soften that up and get it to work with continuums as well. So I'm actually using a mixture of orchestral and synthesised samples, but then convolving samples with tanpura drones at different pitches, even pitches that are a semitone apart. So a flute might be going through a tanpura tuned to C, but then the pizzicato violin sounds might be going through a tanpura tuned to C sharp so you get the filtering based on the tanpura pitch and then the interesting clashes between the different things the tanpura is letting through.

And you do stress that, although there's algorithmic processing taking place in the visual and music construction, the relationships between music and visuals are your input. You take control of that process.

Yes, it's tempting to look for a Holy Grail, to find some way to automatically deal with them both, but the way I phrase it is that what moves us in music is so complex, and multidimensional and temporal in its nature, and the computer knows nothing about that. And the same thing goes for image, what really moves us in an image, what provides a sense of coherence for us, is massively multidimensional, and in a moving image temporal, and the computer knows nothing about that. And most classical algorithmic techniques know nothing about perception. And so I think it's a common trap, but I consider it naïve to think that one can use a one reductionist parameter from one domain and map it to a reductionist parameter in the other domain and expect it to have a deep relationship. So to me, it's still very much an artistic intuitive task to forge the relationship between the two. And for me, that's lots and lots of trial and error. Lots of solutions that aren't quite right, and sometimes when it does work, it's not necessarily easy or obvious to articulate why or how.

Yes, and sometimes it's serendipitous, you try something and it just works, or it doesn't work as well, but I think to a certain extent you can put any sound to any picture and something will happen and it's your intuition that decides which bits of those are of most interest and you decide to choose and build on them. Estuaries is coming to a conclusion now. Are you going to summarise or consolidate the work in that series?

Part 1 through 4 is done. I haven't seen it presented as a continuum as a multi movement whole yet, which will be interesting to see if and when that happens. This piece I'm working on now is kind of using the same technique. So I have this installation, *Three Breaths in Empty Space* (Battey 2019), which is a very ambient, contemplative piece using the same technology I use for Estuaries. And that's what I'm using for this new installation I'm working on now, again a very contemplative piece, and I might be going more in that direction. But yes, I suspect, for the reasons we described, that I may end doing the plugins. The question here is also real-time versus non-real-time. I have a few live performance pieces I've done the last few years and I have just not been satisfied with the results there, so every once in a while I go that way, explore and then go back to where I really feel like I'm getting my best results with the non-real-time stuff.

What are the key issues, when you take it into live performance, that make it more uncomfortable?

I'm most satisfied when I really have meticulous control over time. It's very hard to find paradigms that give you truly meticulous control over time, audiovisual, in the live performance approach.

I'm trying to move towards live, partly because of not having the tools to create fixed media that I used to have, but I'm enjoying it. I also feel there's an element of performance in composition as well, although it's in a different way, and performance is a way of generating ideas for fixed media.

Yes, the algorithms I'm using and the music, the variable coupled map networks, is a real-time process and I have my controller board with all these knobs and I sit and I just improvise with it until I start to get a sense of its behavioural potentials and gradually I create a structured improvisation around that, record a few takes, then that then becomes the start for the music work and then I edit from there. But given the slowness of the visual algorithms I use, that's been less used. But I did use Open Frameworks to create a real-time version of my Brownian Doughnut Warper algorithm and did use that for two live pieces. *Triptych Unfolding* (Battey 2014), which I was satisfied with, I thought that worked lovely. It's got to be performed as a chamber music piece. So a pianist, me performing the visuals and Hugi Guðmundsson was the composer in real-time, manipulating electronic sound. We were able to watch each other, take cues off each other, stretch the time and really do it like a chamber performance. And I found that quite satisfying, Difficult to create but yes, satisfying.

Is there an aspect to the collaboration that pushed you a little bit harder and in a different way?

Yes. The collaborations up to this point tend to come through commissions. So that was a commissioned work where we were invited to work with Hugi. I was invited by the Reykjavik Centre for Visual Music. So that pushed me in a way that I wouldn't have gone otherwise. The same thing with *The Five* (Battey 2015), which was for vibraphone and real-time visuals. A percussionist had contacted me and asked me to make a piece for him. Andrew Spencer was the percussionist.

And is collaboration something you think you are interested in pursuing?

Yes, collaboration is on my mind right now. I'm feeling like a whole long thread of work has come to a conclusion after almost 20 years now and need to find a new direction. So part of what I'm asking is whether collaboration needs to be part of that. So a different kind of relationship to people perhaps, both as audiences and as co-creators. It's getting harder and harder to find satisfaction, sitting and banging my head against code for days and weeks.

If somebody wanted to learn about your work, what is the piece that summarises the last 20 years of your work most? Is that even possible to list?

Of course I can cheat by referring to series rather than individual pieces too. So I would say the Luna series, Clonal Colonies series and the Estuaries series, that's really been the core of the work. A piece to me that I think speaks to me still particularly deeply would be *Autarkeia Aggregatum*. That one still, I think, achieved something that I find particularly special. *Lacus Temporis (Lake of Time)* (Battey 2008) may not be highly representative, but I still feel like it is a very deep piece that hasn't received perhaps the recognition that it deserved. But it's very slow-breathed and subtle through some of it. I think *Sinus Aestum* (2009) overall is a dramatic shaping and statement and probably does a good job of saying something essential about my language.

Good choices. And in terms of written work?

The 2020 article 'Technique and Audiovisual Counterpoint in the Estuaries Series' (Battey 2020) probably is the most fully developed item talking about one line of thought. Other than that 'Creative Computing and the Generative Artist' (Battey 2016) is going to give the most insight to the work from *Autarkeia Aggregatum* up to just before the Estuaries series.

You start off with a very strong concept, I think. How much does that evolve as things start to emerge from the process? Do you stick rigidly to that original idea?

I tend to see the process as more exploring materials rather than starting with a particular preconceived concept. So I'm creating the algorithms image-wise or sound-wise. I explore with them, I dialogue with them, and start to get a sense of what they say to me and I'm looking for something that speaks to me strongly. Music usually precedes image in most of my work. I'm hunting for these musical signatures that move me in some way and then I'll start developing those and I'm looking and asking what does this mean, what's the energy under it? And sometimes the concept will then come out of that. I can decide things, but even then, because the nature of my tools, both image and sound side, they're not ones where I can say 'I want this' and go create it, I have to dialogue with those tools and find out what they can do, which is sometimes, or even often not exactly, what I think I want it to do. So through that dialogue, gradually something starts to take shape. It emerges out of a dialogue with the material and processes less than being defined. And so I might think this needs to build up like this a few ways and come to a climax here. I might have that kind of concept.

So there is a particular quality and flow to the music? When you then start working on the visuals are you looking at those musical qualities in the visuals or is it more about the visuals just having their own complementary aesthetic?

One way I've sometimes framed it is I want it to seem that the image and sound together are the expression of some meaning-complex that is not necessarily either musical or visual in its nature. Somehow they're both expressions, maybe in a very parallel way, maybe in a somewhat divergent way, but are expressions of some coherent energy underneath.

So it's a relationship between them, not necessarily the individual elements, it's the cohesion between them?

Yes, and it's a holistic cohesion. That's where you have to back up and just let perception fall open and say this visual texture and this musical texture here almost seem unified but not quite. The musical material has more of a sense of randomness in it, but the image is too ordered here. Maybe we need to figure out some way to create a little bit more disorder in that image and then, there we go, now they feel like they are coherent expressions of a common energy. That's the kind of thing I'm looking for. With something like *Estuaries 3* (Battey 2018) I was pushing myself to look more towards this idea of interdependence of materials. Materials are not necessarily just holistically synced, but, have a little bit more of a fluid counterpoint relationship to each other. You mentioned sound, but they still feel coherently like they're coming out of an underlying energy.

Is there any influence which comes from psychological concepts? I know you've mentioned Gestalt theories in the past, but also meditation. Do they inform what you do?

I think the meditation informed my sensibility, the certain kinds of things I'm drawn to or not. I find it a little bit paradoxical or strange in the sense that even though in normal life I'm a fairly quiet and reserved person and you think OK, meditation fits that, but in fact I create quite dramatic works.

I feel one of my greatest strengths is the ability to shape this coherent, dramatic statement over time with these abstract sound and image materials. But sometimes I wonder, how does that relate to the relative detachment one often has through meditation. But of course reality is that insight meditation is about opening oneself to whatever is arising, including the emotional rather than walling it off. So in some sense, I become a more emotionally open person through the meditative process. Maybe in some way the music itself has always been an aspiration towards that as well. Balance off my very calm, reticent, reserved public life with very dramatic musical expression.

Yes, that reservedness needs an outlet in some way. It is a form of expression ultimately.

Yes. On the question of Gestalt psychology, it's more that I went to Gestalt when I was trying to find some explanation of what I was already doing. The phrase I came up with early on was trying to say 'what makes this tick'. And I came up with this convoluted phrase of 'isomorphism of complex gestalts'. It's in an unpublished paper called 'Isomorphism of Complex Gestalts: the Audio Visual Composition *Autarkeia Aggregatum*' (Battey 2010). It was from the Visible Sounds Conference at Hebrew University in 2010. Little fragments of that made their way into other articles, but there I was trying to somehow point to this idea of a holistic perception, something that even if we break down the sound into its little parts or its notes, and even if we track every particle on the screen, that's irrelevant. The objective is irrelevant. What's necessary is that phenomenological immersion in how these two things are working together. And perhaps this relates to meditation too, I say when I'm trying to get these things to work together I'm listening to my whole body-mind. What is the whole body feeling in response to this image flow? What is the whole body feeling in response to this sound flow?

References

Battey, Bret. 1997. *On the Presence of Water (1997)*. Audiovisual Composition. https://vimeo.com/34502481.

———. 2004. 'CMatrix10 / CMatrix12'. Audiovisual Composition. Bret Battey—Gallery. http://bbattey.dmu.ac.uk/Gallery/cmatrix10.html.

———. 2005. *Autarkeia Aggregatum*. Audiovisual Composition. https://vimeo.com/14923910.

———. 2008. *Lacus Temporis (Luna Series #2)*. Audiovisual Composition. https://vimeo.com/13387871.

———. 2009. *Sinus Aestum (Luna Series #3)*. Audiovisual Composition. https://vimeo.com/12854049.

———. 2010. 'Isomorphism of Complex Gestalts: The Audio-Visual Composition *Autarkeia Aggregatum*'. Paper presented at Visible Sounds: Interrelationships among Music, the Visual Arts and the Performing Arts. Hebrew University, Jerusalem., February, 2010.

———. 2011. *Clonal Colonies: I—Fresh Runners*. Audiovisual Composition. https://vimeo.com/32143877.

———. 2014. *Triptych Unfolding*. Audiovisual Composition. The Reykjavik Center for Visual Music. https://vimeo.com/121612772.

———. 2015. *The Five*. Live Performance. http://bbattey.dmu.ac.uk/Gallery/thefive.html.

———. 2016. 'Creative Computing and The Generative Artist'. *International Journal of Creative Computing* 1 (2–4): 154–73. https://doi.org/10.1504/IJCRC.2016.076065.

———. 2018. *Estuaries 3*. Audiovisual Composition. https://vimeo.com/264837797.

———. 2019. *Three Breaths in Empty Space*. Audiovisual Composition. Leicester. https://vimeo.com/440246221.

———. 2020. 'Technique and Audiovisual Counterpoint in the Estuaries Series'. In *Sound and Image. Aesthetics and Practices*, edited by Andrew Knight-Hill. Sound Design. London: Focal Press.

———. 2021. *Estuaries 4*. Audiovisual Composition. https://vimeo.com/662716390.

Cope, Nick and Tim Howle. 2003. *Open Circuits*. Electroacoustic Movie. https://vimeo.com/633346.

———. 2006. *Son et Lumieres*. Electroacoustic Movie. https://vimeo.com/633826.

Eboman. n.d. 'EboSuite—Turn Ableton Live into an Audio-Visual Production Suite'. Accessed 16 June 2022. www.ebosuite.com/.

Harris, Louise. 2011. *Fuzee*. Audiovisual Composition. https://vimeo.com/19164744.

———. 2014. 'Ic2–Louise Harris'. *Louise Harris Sound and Audiovisual Artist* (blog). www.louiseharris.co.uk/work/ic2/.

———. 2017. *Alocas*. Audiovisual Composition. www.louiseharris.co.uk/work/alocas/.

———. 2018. 'NoisyMass'. Performance presented at the Sonorities Festival, 21 April. www.louiseharris.co.uk/work/noisymass/.

———. 2021. *Composing Audiovisually: Perspectives on Audiovisual Practices and Relationships*. 1st edition. New York: Focal Press.

———. 2021. 'Filigree Traces–Louise Harris'. *Louise Harris Sound and Audiovisual Artist* (blog). 2021. www.louiseharris.co.uk/work/filigree-traces/.

Ishii, Hiromi. 2013. *Refraction*. Visual Music. https://vimeo.com/59052220.

———. 2018. *Avian*. Visual Music. https://vimeo.com/255059849.

———. 2019. Visual Music. *Kiryu-Sho*. https://vimeo.com/350776853.

———. 2020. *Ice*. Visual Music. https://vimeo.com/474116472.

Levin, Golan. 1998. 'Yellowtail—Interactive Art by Golan Levin and Collaborators'. Flong. www.flong.com/archive/projects/yellowtail/index.html.

Oliveira, João Pedro. 2017. 'Sound, Image and Dance Interaction in Neshamah'. In *Video Dance Studies*. Politecnica de València. https://videodance.blogs.upv.es/iv-meeting-2017-schedule/.

———. 2018a. *Neshamah*. Visual Music. https://vimeo.com/255169571.

———. 2018b. *Tesseract*. Visual Music. https://vimeo.com/272369934.

———. 2019. Visual Music. *Things I Have Seen in My Dreams*. https://vimeo.com/3361 05455.

Playmodes. 2022. 'FORMS–Screen Ensemble |'. FORMS–Screen Ensemble. 2022. www.playmodes.com/home/forms-screen-ensemble/.

Snibbe, Scott. 2002. 'Deep Walls—Interactive Art'. Scott Snibbe. www.snibbe.com/art/deepwalls.

Sweet, Jack. 2020. 'Keysight on Steam'. 2020. https://store.steampowered.com/app/1325 730/Keysight/.

Thompson, Tim. 2010. 'LoopyCam'. Timthompson.Com. https://timthompson.com/tjt/loopycam/.

———. 2012. *Space Palette at Symbiosis 2012*. Symbiosis. www.youtube.com/watch?v= M-Jro6IcAtQ.

———. 2020. 'Space Palette Pro a Visual Music Instrument'. Self published. https://nosu chtim.com/tjt/particles/talks/space_palette_pro_2020.pdf.

———. 2021. 'Space Palette'. Notes from NosuchTim. 6 November 2021. https://nosuchtim. com/spacepalette/.

———. (2021) 2022. 'Space Palette Pro'. Space Palette Pro. G-code. https://github.com/ vizicist/spacepalettepro.

———. n.d. 'Adventures in Casual Instruments and 3D Input'. Self Published.

———. n.d. 'Igesture Pads'. Timthompson.Com. Accessed 16 June 2022. https://timthompson. com/tjt/images/igesture_pads.jpg.

Weinel, Jonathan. 2013. *Mezcal Animations #1–3*. Audiovisual Composition. https://vimeo. com/69790818.

———. 2018. *Inner Sound: Altered States of Consciousness in Electronic Music and Audio-Visual Media*. 1st ed. Oxford, UK: Oxford University Press.

———. 2019. 'Cyberdream VR: Visualizing Rave Music and Vaporwave in Virtual Reality'. In *Proceedings of the 14th International Audio Mostly Conference: A Journey in Sound*, 277–81. Audio Mostly. New York: Association for Computing Machinery. https:// doi.org/10.1145/3356590.3356637.

———. 2021. *Explosions in the Mind: Composing Psychedelic Sounds and Visualisations*. 1st ed. Singapore: Palgrave Macmillan.

3   Perform

Live Electronic Visual Music

Performing Electronic Visual Music

For all artists, including the visual musician, performance involves stepping into the liminal space of the concert venue. Leaving behind the intimate safety of the studio environment presents many uncertainties and a new set of challenges. Each venue will possess its own unique character and ensuring all the equipment works as expected in the new space presents further complications. When the doors open, the audience bring a sense of expectation and excitement elevating the performer to a heightened state of awareness and responsibility. The focus exercised during production and rehearsal is transformed by the impulse of the moment 'like the tightrope walker on the high wire, each move is absolutely spontaneous and part of an endless discipline' (Schechner 2003, 54). At this moment the performer must balance their skills and the performance challenges to create the best possible outcomes. The time-constrained nature of performance presents its own problems. Whereas the studio composer has a wealth of time and tools to perfect their artwork before releasing it into the world, the performer has a finite time and set of constraints with which to express themselves. Timing and synchronisation are critical factors for the visual musician and sound and image relationships, which the composer has precise control over during composition, become a more nebulous and difficult-to-manage prospect in real-time. It is possible to tightly synchronise some events, but more complex, gestural, audio-visual behaviours and relationships can be difficult to reproduce. Nervousness and performance anxieties can also inhibit the performer such that they are incapable of performing the work in its intended form, thereby putting their credibility at risk. This increased potential for problems, reduced control and performance pressures are amongst the reasons that some artists avoid playing live at all.

In visual music performance the technical requirements for playing live are often more complicated than those for the musician. Working in multiple media necessitates the use of a range of visual and audio technologies, creating more opportunities for glitches or, at worst, complete darkness and silence. It only takes one device to fail or behave in an unexpected manner to potentially throw the performer off balance or prevent the show going ahead. Even when everything works as expected there are issues related to the complexities of the performance system and the art it helps create. Through recording technologies, it is of course possible

DOI: 10.4324/b23058-4

to create a 'perfect' version of a performance, the stage and setting can be carefully designed, multiple takes can be recorded and compiled, the audio and video quality manipulated until the best representation of events is ready for release. Despite this, artists still perform in person, and audiences want to see them do this. Why is this the case? A key aspect that keeps surfacing when discussing live performance is the one associated with risk. The risks that must be negotiated include all aspects and outcomes of performance such as equipment failure, unexpected development of the unfolding work and a compromised reputation of the artist (Henke interviewed by Hermes (Hermes 2021, 40–48)).

Despite all of this, venturing onto the stage can be a rewarding experience for those willing to take the risk. Performing live is a direct way for an artist to interact with an audience. During an event there is immediate feedback, contributing to a flow experience, and discussions can flourish afterwards as the performance sparks interest and debate. Creative relationships and collaborations can begin to emerge from these conversations and ideas fomented in the artist's mind. The fixed-media composer on the other hand may have to wait weeks or months until their work is finally revealed, and feedback received sometimes simply as comments on online sharing platforms. Testing out new media materials in a performance space is a good way to quickly test ideas, gauge response and inspire one to continue. The pressures of the performance can be a positive force, disabling the critical function and producing unintended results and even mistakes which on reflection are desirable and make it to the final cut of the fixed-media. There is also the unquantifiable 'buzz', and adrenaline rush mentioned by Harris in Chapter 2, that is experienced before stepping onto the stage, when things go well and which can linger long after the event.

From a fan and audience perspective, the spectacle of watching the live performer is desirable and enjoyable. They want to be close to their heroes or peers, to experience the artist performing their work in a social setting. The concert is a dedicated time and space devoted to appreciation of the artist. Concerts can possess a degree of exclusivity. Venues have a limited capacity and accessibility, there are a limited number of shows. Some performances are remembered as breakthrough or peak moments that people reminisce about. In the post-COVID-19 era, attendance at events is gradually returning to pre-pandemic levels. People are once more venturing out and creating a vibrance that was lacking during multiple lockdowns. The pandemic created more opportunities for attending streamed events and, although these have become more widely accepted, there is still a desire to experience the artist performing their works in a venue through high-quality audio-visual systems.

Liveness and Performance

The ability to experience 'non-live' events is a relatively recent development. Before the recording of performances was possible, all events had to be experienced 'live'; it was necessary to watch performers in person. It was impossible to listen to a recording of these performances because the technologies simply did not exist. This is how the differentiation between 'live' and 'not live' originated, but recent discussion is more nuanced and relates to degrees and qualities of 'liveness'.

Corporeal, interactive and temporal and others are all types of liveness that may, to various degrees, be present. For example, temporal liveness is 'where music is live during the time of its initial utterance' (Sanden 2013, 11). This would not be the case at the streaming premiere of a pre-recorded performance, but this same stream could possess a measure of virtual liveness and be perceived by the audience as 'live'. A visual music event, involving more than one performer, can exhibit interactive liveness as each member of the ensemble takes visible cues from the other. There is also a distinction to be made between the terms live and performance. Live refers to specific qualities of an event described above, whereas performance in its broadest sense relates to motion and the displaying of activities to others. Performance can be observed in many areas of life including ritual, play and politics, but a visual music performance, in Schechner's classification (2003), falls between the categories of art and popular entertainment. Furthermore, a visual music performance can encompass all the functions of performance including a capacity to entertain, to educate, to create beauty (2020, 7–21), and I would extend this to include its ability to provoke thought and novel apprehensions.

Performance and Flow

Does performance require a different focus to composition and how can the energy it demands be channelled productively? The physical act of performance requires gestures, action and concentration under pressure and can cause psychological reactions such as anxiety and arousal. In an ideal scenario a well-rehearsed visual music set will be translated into a free-flowing performance. Positive feedback from the audience will reinforce the quality of the performance creating a self-rewarding reciprocal relationship between audience and performer. Flow is a concept (Csikszentmihalyi 2008) that has been particularly inspiring for me recently as a way of trying to get the most out of an experience, which could involve any activity, but it is useful in this context as it uses strategies that can be applied to performance. The flow state is one in which the sense of time becomes distorted, you are fully immersed in the activity, and it becomes an autotelic experience, one which is rewarding for its own intrinsic value. Flow is achieved by setting clear goals, receiving unambiguous feedback and matching skills with challenge (Csikszentmihalyi, Montijo and Mouton 2018, 220). All these factors are possible during performance. It is however important to match the challenge with the current level of skills; in this situation arousal and control can lead to flow. If, however, there is an imbalance between skills and challenge, negative responses can result. The virtuoso musician, for example, will feel at ease when asked to recite a simple melody, but can become bored or apathetic if this has to be repeated. Conversely, the novice musician will be overwhelmed and anxious if asked to perform a complex piece in front of an expectant audience. By setting the challenge at an appropriate level to one's capabilities, the situation can become more rewarding and manageable. The novice must therefore begin with modest goals and the expert with more lofty aspirations.

In music performance flow has been shown to be achievable in music ensembles more readily than in a general population (Spahn, Krampe and Nusseck 2021). In this situation, the goals are clear, feedback from the ensemble and audience is immediate and the musician is chosen to perform due to their suitable skill level. In the box later in this chapter James Dooley also describes experiencing an intense flow state while performing his music. My own anecdotal experience, however, is that flow is more difficult to achieve in visual music performance than in music performance. I believe this is in part due to the complexity of the performance systems but also the requirement to focus attention on and between multiple media streams. Whereas the audience will fuse the multimedia experience into a cohesive whole, the performer on the other hand, especially the solo visual musician, must constantly switch attention between sound and image from both an aesthetic perspective and a technological one. These difficulties can be minimised by preprograming and preparing materials where possible and using an optimised controller setup with thorough rehearsals. In Chapter 2 Weinel describes the extensive preparation work that goes into producing materials to fill an entire set. This alone poses a greater challenge than producing visual or sonic materials alone. The fact, however, that there are greater challenges in visual music performance also implies that, when one becomes more adept and things go well, higher levels of flow are achievable. This should translate into a more enjoyable experience for performer and audience alike.

Real-time Sound and Image In Context

The intersection between music and visual performance occurred relatively recently. Whereas music performance has a long history, dating back to the development of the earliest musical instruments, visual performance, using technologies described in this text, originates around the late 19th century with the advent of colour organs. Subsequent developments included expanded cinema, liquid light shows, Lumia and, most recently, video projections. This represents a period with rapid technological developments, which has coincided with visuals being increasingly integrated into music performances. Before this, the earliest innovations in live visuals took place in what could be considered a 'silent' period. The original colour organs were silent devices and the advent of experimental film during the late 1920s and early 1930s saw the creation of many silent animations. These early experimental films can be considered as a type of 'fixed media', once recorded onto the film roll the visuals were not modified and were projected in their original form in cinematic settings. There was no capability for performative interaction, apart from control over playback. Soon after these early experiments, sound started being added to film or performed alongside it and the visual music artform was practiced and developed during the following decades as detailed in Chapter 1. During the 1960s the expanded cinema movement took film and music into unconventional venues and combined visual projections with sound to create cohesive multimedia events often inspired by spiritual and metaphysical beliefs. There was a performative aspect to some of these events not experienced with the single-screen-based presentations of visual music films in cinemas. During the 1960s, liquid light shows (Steve Pavlovsky n.d.; Gibson 2022) also began to be incorporated into live music shows combining live sound with 'performed' abstract imagery. The liquid light artist could interact with liquids and inks in response to events happening on stage. In the UK, automated projection devices such as the Optikinetics Solar 250 accompanied discos and band performances during the 1970s (Foakes 2022), producing similar results to liquid light shows, but without the same level of interaction. When cheaper video VHS systems became available during the 1980s they were initially used to store visuals to be projected as fixed media. This opened the visualist to a greater range of materials and an increased level of design of the visual content. From here onwards the rise of the home computer enabled rapid strides in visual creativity. Originally used to generate visuals, which could then be recorded to VHS tape, their rapid advancements meant that real-time processing soon became available, and effects could be added in real-time, allowing for a true live interactive visual performance. This principal technique has continued to date with ever-increasing refinement and power. Recent technological developments have further expanded the possibilities for using visuals alongside live music. For outdoor concerts in unusual venues, high-power projectors permit projection mapping onto large surfaces such as buildings. Display technologies that do not require projection, such as LED walls and stealth screens, can be located in front of or behind the performing artist for a range of creative effects. On a more modest scale, inexpensive, powerful computers have enabled single-screen projections to accompany band or solo performances in small, intimate venues. For the visual music artist these technologies permit simultaneous renditions of audio and visual media and give the performer a means of interacting with their media in real-time.

Visual Music Performance

As hinted above, there are several types of performance involving combined sound and image. *VJing*, *live cinema* and *audio-visual performance* are key practices that share similar traits but involve different performance scenarios, technologies and materials (Gibson and Arisona 2022).

The *VJ* will often improvise using a large library of video clips which are processed by various visual effects in response to the DJ selection. VJs can be further classified as narrative and wallpaper VJs . . .

> The Narrative VJs will emphasise on their content/delivery as the core entertainment from their performance. The audience is expected to "watch" their set and appreciate it on an equal, yet individual level with the DJ (or live set act . . .).
>   The Wallpaper VJs will be more concerned with the overall visual environment and propose content which offer cohesion with the style and layout of the venue and the music policy.
>
> (Bernard 2006)

Both these VJing approaches have merit but in relation to the theme of this book the narrative VJ aligns more closely with the ideals of visual music.

In contrast to the VJ the *audio-visual performer* uses mapping between sound and visual parameters, so the visuals reflect changes in the music and vice versa, thereby creating an intrinsic relationship between media. A variety of controllers can be used to interact with both sound and image and both media can be generated in real-time via hardware and software. The range of music genres used in audio-visual performance is much wider than that heard alongside a VJ.

The unique characteristics of *live cinema* include the use of longer shot sequences and a tendency towards greater narrative. Live cinema materials are often performed using purpose-built rigs and can involve a variety of screening methods including analogue film and projectors.

> The live cinema event at its best promises a much more integrated media experience in which all elements are 'speaking' to each other, in a sense creating a synesthetic experience for the audience, in which all mediums mutually reinforce one another.
>
> (Gibson and Arisona 2022, 305)

Visual music shares many traits with the above practices but there are some key differences. One of these is the way that the visual materials are considered and utilised. The visuals in audio-visual performance will be tightly integrated with the music, but in visual music performance musical qualities will be embedded within the visual medium itself and will influence its performance. In real-time work this will involve a degree of improvisation with visual materials treated as if they are musical materials. This can be achieved by techniques 'such as a structural reference to musical composition, by transcoding sound into image or vice versa, or by performing sound and image according to the rules of (musical) improvisation' (Carvalho and Lund 2015, 35). The relationships between sound and image are also paramount so that 'the central point of visual music is, indeed, the quality of the audio-visual combination . . . The result of this audiovisual combination should be a new, genuinely audiovisual product' (Carvalho and Lund 2015, 35). The intent to create synergy between both media must be the goal of the visual music performer more so than the alternative approaches described above. In some scenarios, particularly with the wallpaper VJ, the visuals can be a secondary concern for both the musicians and audience alike, they are considered as an accompaniment to the music concert. In a visual music sense this chapter is intended for those desiring to perform music and image where both media will share equal importance and qualitative traits. This type of visual music performance can be achieved by a single artist creating and interacting with both media simultaneously, or by a group of artists each taking a discrete responsibility on a specific aspect but working in unison towards a shared visual music ideal; a true synergistic visual music experience.

Characterising Visual Types

The visuals used in live visual music performance can be broadly classified as being either artistic or informative. Most visuals used in this situation are artistic in nature, but informative-type visuals can play a role in performance.

Artistic Visuals

Artistic visuals are the principal type to be experienced in a visual music context. Artistic in this sense means expressive visuals that attempt to convey a coherent aesthetic shared between sound and image. These types of visuals will have a synergistic effect when coupled with music which can be enhanced through techniques of synchronisation, gestural congruency and other qualities as described in the next chapter. They should embody musical qualities.

Abstraction in Artistic Visuals

In artistic visuals, musical traits such as gesture, texture, dynamics and tempo can be expressed in visual form by various means. A common approach to create these traits is to use abstract animation, as it gives the artist detailed control over the visual's behaviour. This does not, however, exclude the use of film footage of real scenes. Film footage can still be abstracted, edited and treated in ways to make it more 'musical' and synergetic with the soundtrack. Abstraction, as a process, removes the source of causation of the visual figures, similar to how spectromorphological sound processes create surrogacy in recordings. This allows the mind to focus on the behaviour of the image itself, and its interaction with sound, rather than its recognised qualities prompting extrinsic associations.

Visual Illumination

The illuminative quality of abstract imagery is also an important factor in its use. All video projections will illuminate the space they inhabit but, when using abstract shapes and blocks of colour, the lighting effect can be emphasised and used as a creative tool. Watch the brief sequence between 9:44 and 9:57 of Alva Noto's performance in Copenhagen (Manchine 2016) to see how the projected visuals illuminate the room and audience. Furthermore, if visuals are projected onto walls and other surfaces in a room, the figures contained in the projection become more distorted and abstracted by the architecture. Projection mapping is one method of capitalising on this effect and can be so convincing as to make it appear that the building or surface itself is moving and changing shape. The Sydney Opera House Facade projection (URBANSCREEN 2012) is a good example of this. Lower-powered projectors can still produce useable results when projecting visuals onto walls, as was done in the Middleport performance of Heatwork (Estibeiro and Payling 2020). In these situations, the projections themselves become a light source and added layer of interest not present when displayed on a standard screen. This illuminative effect, as a by-product of the main projection, should be a consideration when designing a visual instrument and performance environment and can be incorporated into their design. Large expanses of colour, for example, can be generated by an instrument rather than intricate gestural figures. Shifting between different colours can create dramatic effects and change the atmosphere of the space.

Light-based Artistic Visuals

Light-based visuals can be considered a sub-category of artistic visuals.[1] Lumia and liquid light shows fall into this category and are created from dedicated lighting sources often filtered in some way to enhance various qualities. Lumia have very expressive and ethereal qualities with related parallels in music. See my discussion in the next chapter and Keely Orgeman's book (2017) on Lumia for more details. A more intense lighting effect can be produced by a stroboscope. A strobe produces a visually rhythmic effect by producing rapid flashes of high-intensity white light alternated with no light. In an otherwise darkened space, this flashing activates an almost visceral response. Similar results can be achieved with a projected video by alternating between pure black and white video frames. Tempo adjustments can be created by using different frame rates. The structuralist film movement (Moolhuijsen 2021) made use of this technique, the film *Flicker* by Tony Conrad (1966) being a stark example. More subtle lighting effects can be achieved in projections by gradually fading between colours rather than step-change intensities alternated frame by frame. The beta and phi phenomena (Steinman, Pizlo and Pizlo 2000) are another type of kinetic lighting effect that can create the impression of motion in a fixed lighting system and can also reflect rhythmic musical qualities.

Informative Visuals

Informative visuals are the second primary type of visuals, and these can communicate to the audience various aspects of the composition, interactive system or the performer's activities. One limitation of computer-based performance is the inability of the audience to equate the gestures of the performer with the media being created. As discussed later, there may be no obvious relationship between what is performed and what is perceived. This can be alleviated in part by positioning the performer so the audience can observe their interaction with the system and is further enhanced by the performer using more gesturally dependent control interfaces with increased performative gestures. For further visual enhancement a projection of the performer themselves, and/or their interaction with the equipment, can be incorporated in the main visual projection; this would then be an informative visual. Similarly, a video feed of the performer could be displayed on a separate screen or monitor, so as not to interfere or distract from the main visuals. During the first live performance of *Biphase* (Payling and Dooley 2022a), the music was performed by James Dooley from a remote location. To make the audience aware of James' interaction a live video stream, from Zoom, was displayed on a separate laptop facing the audience and next to the author performing the visuals. This reinforced the fact that two people were performing the set. As well as this, some performers may choose to project some components of their performance patch, maybe in a stylised way, to give some perspective of the mechanisms being used. An example of this approach can be seen in Si Waite's *Brains Need Bodies Too* (Trevor 2018). Informative visuals like these are not implicitly necessary for visual music performance but can be used as above and for similar purposes.

The Stage and Setting and Audio-visual Space

In visual music, and other audio-visual performances, there is huge potential for organisation and manipulation of the environment. Some of the key decisions, such as the location of the screen and audience, might have limited flexibility but there are abundant possibilities for positioning of the performer, multiple projections or lights and placement of loudspeakers, all of which can affect and enhance the quality of the environment and experience. A typical configuration for performance with visuals is the use of a single-screen projection with the performers in front or off to the side. Some visualists may also face the screen and be seated in the middle of the audience. These approaches work successfully for the most part but, when on stage, the performer may obscure some of the visuals. With single-screen projections there can also be a tendency for the audience to be absorbed by and drawn to focus on the screen at the expense of the performers. This could be a response learned at the cinema which demands concentration directed towards the screen. Of course, this could be the intention but there are alternatives, such as the use of multiple projectors. By introducing 'polyspatiality' (Ciciliani, Lüneburg and Pirchner 2021, 36), the technique of using two or more projections, the 'pull' of the screen is reduced by creating several focal points. The audience can then choose to focus on different elements throughout the event, creating a more flexible experience. This correlates with my own experiences using television sets either side of the stage as discussed in the final chapter. It put the band centre stage with the visuals contributing lighting and additional visual enhancement. The audience's eyes could wander as desired between each visual element.

Another perspective is to consider the overall visual music concert experience as being created by a combination of the physical space inhabited by performers and audience and the 'virtual' space created by the visuals. Sá and Tanaka (2022, 256) describe three levels of 'arena', progressing from local to distributed and finally the extended arena. The local arena is that around the performer, which is often hidden from the audience and is where the performance itself takes place. There are ways that this local arena can be made more visible such as by projecting the performer's actions onto a screen or by positioning as described above. While some (Joaquim and Barbosa, 2013) advise against using the laptop for illumination and flashing lights in laptop music performance, when working with visuals, light from the instrument can be a contributory factor to enhancing the audience's experience of the event. It brings the light into performer's 'local arena' (Sá and Tanaka 2022, 256), bringing them into focus and integrating them into the visual scene and distributed arena. The distributed arena is the space in which the performance is being held. This is the performance venue which can be designed and manipulated by the positioning of stage and audience, lighting effects, polyspatiality and so on. The layout of this is often limited by the design and architecture of the venue, but there could be some opportunities for creative placement of people and technologies. Beyond the distributed arena lies the extended arena, whereby it is possible to create the illusion of being in a space outside the venue itself. This can be created by the quality of the visuals where the 'video can emulate how we naturally see the world. That enables a subjective sense of presence beyond the physical performance space' (Sá and Tanaka 2022, 257). More abstract and synthetic types of visuals can also project the viewer into a virtual space, particularly if large venues and projections are used such as with fulldome screenings. The key is that the video projections are so absorbing and immersive that the audience feels transported beyond their physical location into an extended arena. The visual music event then becomes a multifaceted experience of being physically present in the performance environment but subjectively present in an imagined 'virtual space'. Adding a sonic layer to this already complex sensory experience creates further ambiguities in the relationships between performer, audience and the visual and auditory spaces. These are all aspects that can be manipulated and used to creative effect when designing the performance materials and the performance space. Despite being subject to multimodal sensory stimulation, the audience is able to resolve all the contradictory and complimentary elements into a cohesive whole (Harris 2016), and likely everyone experiences these type of events in subtly or profoundly different ways. The visual music performance environment is therefore ripe for exploration and creative manipulation of sound, light and space.

Digital Visual Instruments

This section introduces the digital visual instrument (DVI), the optical equivalent to the digital music instrument (DMI). Digital musical instruments are discussed in depth in other places, including the well-established NIME conference, so this section will focus mostly on DVIs. The design of DMIs, however, still has relevance and will help inform this discussion. A digital musical instrument requires two main components, a gestural control surface, which captures the physical gestures of the performer, and a sound generation unit, which produces the sonic output (Miranda and Wanderley 2006, 3). In DVIs the gestural input is maintained, but an image generation unit is the second component.

The control surface requires performance gestures to be mapped to changes in the sound or image generator. This requires a mapping strategy involving *parametric mapping*, whereby a parameter on the control surface is mapped to an action in the instrument and change in the output. To convert the performer's gestures into a useable form may require *parameter conditioning* undertaken by a process in the digital instrument. See the following box for some examples of this. These techniques are necessary for independent DMIs and DVIs but mapping in controllers for simultaneous interaction with sound and video creates additional complexity. In this situation, it must be decided how changes in music are qualitatively mirrored by changes in the visuals and which controller element is suitable for this task. Max Cooper (2017) discussed this in relation to mapping audio effects to corresponding visual effects. He suggests a nonlinear relationship between audio and visual parameters to create the most convincing cross-modal interaction (watch his question response from 1:08:00). When designing and programming the instruments themselves, decisions must be made on how to restrict their capabilities. Because DMIs and DVIs are self-designed and/or programmed, and have potentially limitless features, it can be tempting to continually develop them to accommodate every possibility. In practice there should come a point where the development of the instrument is ended, and the utilisation and rehearsal phases begin. It is often necessary for the performer to work around the limitations and constraints of the instrument and develop intuitive interactions with its interface. This is true even of traditional acoustic instruments. Outsiders may expect electronic instruments, particularly computer-based devices, to be easy to play, however, all instruments require a degree of practise and competency to become familiar with all their intricacies and nuances.

Parameter Conditioning

In human computer interaction with visual music instruments there often needs to be some adjustment made to map the performative gesture more intuitively with the audio-visual result. In general use, raw sensor and controller data or audio waveforms are frequently unpredictable, too dense or have rapid variation, making them unsuitable to be mapped directly to a visual quality. When observing the waveform from a microphone recording of a voice, for example, there will be large and rapid changes in amplitude. Mapping these to something that requires more gradual variation can produce undesirable results. For example, if this was mapped to the opacity of a video layer, it would rapidly flicker. Another example is when a master level fader is adjusted to gradually dim the visuals, the reduction in brightness may not be as smooth as desired and can be especially noticeable when fading from a very bright light to complete black. A simple adjustment to resolve both these issues could be to apply a smoothing function such as a filter that integrates the change, removing rapid changes and creating a more natural fade and opacity variation.

A technique for modifying raw data in this manner is *parameter conditioning*, a term borrowed from the electronics known in that discipline as 'signal conditioning'. Signal conditioning involves processes such as amplification, level translation and linearisation (Kester 1999). Often the raw electronic signals are contaminated by noise or at an unsuitable level to be interfaced with other components later in the circuit. They need to be conditioned, that is, adapted to be more useable. Parameter conditioning is used for a similar purpose, to modify the parameters received from the controller device or audio input so they will perform the necessary function more effectively. Parameter conditioning falls into three broad categories, scaling, smoothing and transforming as illustrated in Figure 3.1.

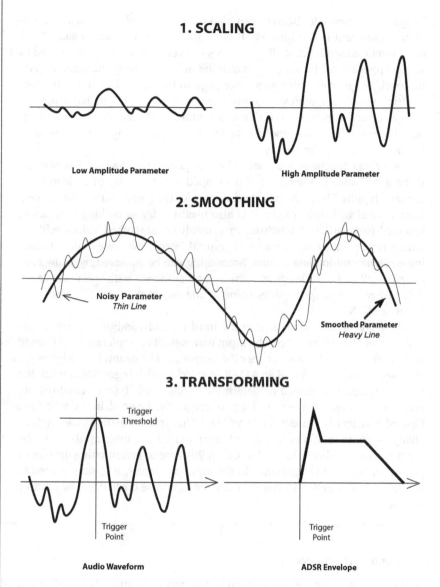

1. SCALING

Low Amplitude Parameter

High Amplitude Parameter

2. SMOOTHING

Noisy Parameter
Thin Line

Smoothed Parameter
Heavy Line

3. TRANSFORMING

Trigger
Threshold

Trigger
Point

Trigger
Point

Audio Waveform

ADSR Envelope

Figure 3.1 Types of Parameter Conditioning: 1. Scale from Low Gain to High Gain, 2. Smooth from Noisy to Clean and 3. Transform from Waveform to Envelope

Types of Parameter Conditioning

Scaling functions are relatively simple and are used where a low amplitude source, for example an audio waveform, is adjusted by a gain control to increase its amplitude. Scaling can also be used to reduce values. A common requirement is to adjust MIDI controller values between 0 and 127 to the normalised spatial co-ordinate values between 1.0 and -1.0 often used in visual applications. In Max this would be achieved using the *scale* object. In TouchDesigner the third parameter page in the *maths* CHOP has a range function which can achieve a similar result. Scaling also includes actions such as inversion between positive and negative and offsetting and rectification, for example when a bipolar (positive and negative) signal needs rectifying to become unipolar positive.

Smoothing functions are used where a parameter changes too rapidly. If the noisy value in Figure 3.1 was mapped to the opacity of a video layer it would rapidly flicker. A smoothed version of the parameter would create a more natural variation in light. It is also useful to try smoothing parameters received from MIDI or touchscreen controllers. Smoothed values will reduce jerky or sudden movements in gestural interaction, potentially producing more natural-looking results. Smoothing can be achieved by a low pass *filter* CHOP in TouchDesigner, 'ease'-type keyframes with spline curves in After Effects or ramped values using a *line* object with suitable transition times in MAX.

The *transformations* created in the third type of conditioning are a special type of conversion where the input parameter is completely transformed into another type of value for specific purposes. The example in Figure 3.1 demonstrates the creation of an ADSR envelope which is generated when the audio waveform exceeds a predetermined value. This type of conditioning could be used on a percussion loop to trigger the layer visibility when the desired transient is present for example. Other transformations can include changing the state of a switch between on and off, creating a pulse, triggering a random number value and so on. In this type of conditioning the output parameter bears no resemblance to the input parameter, it is simply used to determine when the transformed parameter should be created and/or for how long it should last.

Gesture and Performance

A control surface, or controller, can capture a variety of gestures that are determined by the type of device being used. Common controllers include MIDI devices, from manufacturers such as AKAI and Novation, and iOS or Android touchscreen tablets with custom-designed control layouts. Alternate controllers are also possible by using pressure, motion, bend and other sensors (Bongers 2000) embedded in a variety of hardware devices. Interaction with these types of controllers is mostly a tactile process but semiotic gestures can be captured by motion tracking systems which require no contact. A decision must be made about the suitability of the controller and its effectiveness in a performance scenario. This is, at least in part, determined by practical considerations related to functionality, but aesthetics and perception can also be a consideration. Consider traditional acoustic musical instruments, where there is an intrinsic relationship between the interactive gesture and the resultant sound. The physical gesture of striking a real cymbal with a drumstick, for example, would produce a loud ringing onset transient followed by slowly decaying, high-frequency noise. There is a clear, observed, and audible causal relationship between the performed gesture and the sound that is created. Although these performative gestures are necessary to produce the sound, they can also help an audience understand and appreciate the spectacle of the performance. Studies of acoustic instrument performances have shown that they are evaluated predominantly by visual information (Tsay 2013; Vuoskoski et al. 2016). A performance can be judged as being of high quality purely by positive visual information; or to put it another way, even if the music being created is not particularly good, an audience can be swayed by an engaging and convincing performance. Visual gestures are therefore key to the perception of a good performance. The potential for this is somewhat broken down when the instrument is a computer coupled with a non-acoustic controller. In this scenario, typical with DMIs and DVIs, the relationship between gesture and output becomes a more abstract concept. A gestural input for a DMI could be a simple push of a button which in turn triggers a drum loop sample. For a real drummer to recreate that same loop would require them to perform extensive physical gestures. The same push of a button could also trigger any number of recorded samples with no intrinsic relationship between the gesture and the sonic result. Despite this it is possible to design or choose a controller that requires increased or intuitive gestures. The gestures themselves can also be exaggerated by the performer and additional body movements used to demonstrate their enjoyment and engagement. This is not always necessary, however, sometimes a relatively stationary performer can complement the aesthetic of the event. The important factor is ensuring the audience perceives the performer to be masterful in using their instruments, and this ultimately is enhanced by experience and practice, in turn resulting in high-quality visual music.

Biphase Electronic Visual Music Performance

This chapter will conclude with a case study of *Biphase* (Payling and Dooley 2022a), a duo comprising James Dooley (formuls) and Dave Payling, the author. James performs electronic music, using his bespoke music software, and Dave performs visuals created in TouchDesigner. It is electronic visual music in performance form. Some context is given here to help the reader trace the development and technical implementation of *Biphase*. During the COVID-19 pandemic in 2020 many performances, including those from well-known commercial artists, took place online, and it was at this time the *Biphase* collaboration began. Restrictions in place at the time meant that some means had to be found to rehearse and perform in separate, remote locations. The decision was made for James to stream his audio over the internet from his own studio and for Dave to record and stream the combined sound and image for streaming. This approach eliminated any latency between the sound being heard and image being displayed, which would have been the case if they were streamed separately. Fortunately, by the time rehearsals were complete it was possible to perform in person, but James was not able to attend on the date of the first concert. This meant the same technical approach described above was still used, but the audio and visuals could be screened in the venue at NoiseFloor 2022 (Payling n.d.). *Biphase* was later performed with both artists in the same venue at ICMC 2022 (Payling and Dooley 2022b). The advantage of the in-person event was that by having both performers in the same space allowed subtle clues to be communicated when changes were expected or planned.

James' audio software involves creating custom externals involving sound synthesis in PD and controlling all parameters via touchscreen interfaces and a custom interface created in 'open stage control'[2] (Doucet n.d.). For the networked performance James' remote audio performance was streamed to Dave in the venue via Sonobus (Sonobus n.d.), a free, high-quality network audio streaming platform. The received audio was treated in Ableton Live to adjust level and limit the output to ensure it was of suitable quality for the venue's sound system and for patching into the TouchDesigner visual software. Although multiple audio streams from different layers could have been used, giving more options in mapping, it was decided to map a single master audio output to various parameters as described later. One reason for this was to minimise the options for audio reactivity and force the visualist to interact more with the visual instrument. The inclusion of some audio reactivity created the sense that there was a parametrically mapped relationship between the sonic and visual events, but the intention was to create a more performer enhanced result.

I asked James (formuls) about the working methods he uses in performance and composition. The questions and responses are included here:

Dave: Why do you use or prefer to use touchscreens when interacting with your audio patches?

James: To answer this question, it's worth giving a little historical context to the formuls project. I had been trying for some time (around two years) to develop a system with Pure Data that could be used for live electronic performance, something that allowed fine timbral control and temporal independence of multiple sound streams. Importantly, this system should not rely on presets, rather the performer crafted the sounds they wanted in the moment during performance. In 2014 the first formuls system was created using Pure Data. The Pure Data patch contained an arrangement of GUI objects (XY pads, sliders, switches and buttons) in multiple windows to control sound synthesis. While initial tests were performed on a non-touchscreen Linux laptop, this quickly changed to a touchscreen laptop. From this, three Lenovo Android tablets were then used to control the Pure Data patch using a custom TouchOSC interface. During these early stages, the TouchOSC interface went through many layout changes as the system developed through an iterative design process. Choosing a touchscreen interface meant that this type of interface development was possible, allowing for bespoke designs to be tried and tested rapidly and with relative ease. The touchscreen interface has since moved to the open-source software Open Stage Control, though the fundamental GUI widgets—xy pads, sliders, switches and buttons—remain the same, as does the approach to having multiple interface pages that each control a different sound synth.

While a wealth of pre-existing controllers is available, these typically lock the user into a pre-defined interface, which in the case of formuls was not appropriate. Commercial alternatives such as the Sensel Morph, Roli Lightpad and Yaeltex allow a level of customisation in the interface, though only the Sensel Morph and Roli Lightpad provide access to an xy pad. Alternatively, a completely bespoke approach might have been explored using technologies such as an Arduino combined with interface components. Unlike a touchscreen, these tactile interfaces provide the user with haptic feedback, and this was a known trade-off when making the decision not to pursue this route for the control interface, though they are, however, generally a more costly option. With the exception of the Sensel Morph, they are harder to reconfigure, and as commercial products are exposed to the vagaries of the market: the Sensel Morph and Roli Lightpad are now discontinued, which poses sustainability issues. Choosing the touchscreen meant that interface changes could be quickly made, which allowed for experimentation as well as rapid updates to the design.

Open Stage Control is a multiplatform software that runs on a host computer and is accessed via a web browser on the touchscreen device. Consequently, this means the formuls GUI is hardware agnostic as any tablet connected to the same local network as the host computer can load and interact with the interface. Additionally, using the OSC protocol, as opposed to MIDI, allows for higher message resolution. This is particularly useful for controlling the frequency of an oscillator or a filter or making subtle time adjustments to a delay, for example. Fundamentally, the level of customisation that a touchscreen interface can offer is at the heart of why I prefer to use them. As such, the visual design of the interface allows multiple control layers to be presented through modal buttons that activate and deactivate certain levels of parameter functionality while always being able to see the parameter settings clearly. Additionally, the use of XY pads to facilitate the control of two coupled sound parameters through one interaction point adds a level of efficient and expressive control. The formuls interface uses ten XY pads on each of the main synth pages. This configuration allows up to 20 parameters to be controlled simultaneously and independently through the movement of the fingers and thumbs on both hands.

Can you provide a brief overview of your audio patch?

The audio patch can be split into two main sections: six synthesisers with audio processing effects, and master bus processing. The synths are Pure Data externals that were created with faust. The original formuls system used PD native objects to perform the synthesis, though as the complexity of the patch increased, performance issues ensued. Using faust coded externals offered the ability to increase the complexity of the synthesis and audio processing available when creating sound with the formuls system. Each synth instance is a two-operator FM synth, where frequency can be additionally modulated by an incoming signal from another synth or a filtered noise source. By default the synth is in monophonic mode without an ADSR envelope applied. With an ADSR envelope, the synth can be polyphonic. Oscillator frequency and waveshape can be controlled, and a 'slide' parameter allows a sequencer-triggered glide between a start and stop frequency. The range and time of the glide can be controlled. Each synth then allows amplitude modulation, chorus, saturation, high and low pass resonant filters, gate, delay and reverb to be applied to the oscillator(s) output. If the ADSR envelope is engaged, envelope shape can be controlled and a note is triggered by a sequencer. This sequencer can have up to 64 steps, can perform a range of rhythmic subdivisions and has a number of generative options. Finally, each synth allows loudness and panning control, and can receive signals from other synth instances that control an amplitude follower that in turn control the loudness of the current synth.

The audio processing on the master channel is as follows: pseudo-Paul stretch of the incoming audio signal, a 'repeater' effect that records and loops short samples to create glitch effects, a 'digitaliser' effect that combines upwards pitch shift followed by bit reduction followed by complimentary downwards pitch shift, and a brick wall limiter. The pseudo-Paul stretch and repeater effects allow individual synth instances to be sent to them, while the digitaliser and limiter effects are applied to all instances.

How did you determine and what is the musical structure used in Biphase?

The structure is based on a previous composition that was performed online at Noisefloor Festival 2021. This centres on the idea of having a slowing evolving drone layered with other sounds that hint at musical material. This eventually breaks free and introduces more rhythmical elements that explore the ideas previously heard in greater depth. The pace at which this happens was partly a result of giving time and space to the visuals so as to allow them to develop.

The approach is greatly influenced by my studio work, where in albums such as *beforethetitrationofpanacea* (formuls 2018) and *afterthedeathofego* (formuls 2019), the tracks of the albums were created by deconstructing and extrapolating material from an improvisation. The idea is that a limited amount of musical material is reworked to its logical conclusion. There is an emergent coherency between all the tracks of the album due to the sonic limitations of the initial source as well its stylistic elements that influence the construction of the other tracks.

In live performance this is more difficult to do as crafting a particular sound or musical idea is not necessarily an immediate process, like playing a chord on a piano is for instance. Deconstructing and reworking that idea throughout a performance is hard as a result of not necessarily having access to the same tools live as I would in the studio. This is partly what my research aims to address. Consequently, designing a digital musical instrument with an interface that enables the performer to play the sound or musical idea they want as quickly and accurately as possible is important for me artistically, and one that allows me to realise the musical structures and compositional Ideas that I want to.

In what ways, if any, does the visual inform the musical?

Though I was aware of some of the visual patterns that had been experimented with, the main influence this had was to compose music that provided enough time for ideas to sit and develop. In doing this, my intention was that this would give the visual performer time and space to explore different intricacies. The sounds chosen were not specifically influenced by the visuals. Rather, these were chosen from my own aesthetic pre-disposition. During performance, watching the visuals at times informed how quickly I moved through sections. However, during the actual performance watching the visuals was not possible due to them being projected behind me and the intense flow state that I typically experience when performing. I find it very hard to shift my attention away from enacting my own musical ideas when I am in this mindset.

Biphase Visual Design Concept

The *Biphase* visuals were intended to interpret and cohere with the music through intuitive interaction and parametric mapping. This was achieved by designing the DVI with a minimal aesthete of simple geometric forms with some colour manipulation. Although possessing only a few visual elements, there were still plenty of opportunities for parametric mapping and intuitive HCI performance to achieve audio-visual cohesion. Furthermore, although additional enhancement could have been created through the application of post-processing effects, by using Resolume, for example, it was decided to reduce complexity by limiting the performance tool to a single platform, TouchDesigner. This minimal approach to the technical realisation was perceived to complement the aesthetic minimalism. The fundamental building block of the DVI in *Biphase* is a geometric shape that can be repeated and distributed in three-dimensional space. This shape is selected from one of three primitive types, a circle, a torus and a box. Colour and noise effects, created within TouchDesigner, permit additional enhancements and distortion of the shapes. There was a desire to incorporate a strong gestural element in the visuals and this was afforded by the abilities both through parametric mapping and human interaction to move, grow and shrink the shapes, as well as allowing them to move and resize. Shapes moving and morphing in unison to developments in the soundtrack create cohesive gestural relationships between sound and image. As a musical arpeggio is heard, a multitude of shapes moves rhythmically around the screen, for example. A textural visual quality can also be created by distorting the shapes into more abstract forms. When the shapes are distorted by noise they become less defined and more textural in nature. These textural qualities are further emphasised by adjusting light intensity and colour and complementing the tonal and pitched qualities in the music.

Biphase Digital Visual Instrument (DVI) Design

The visuals created by the DVI designed for *Biphase* comprised of two principal components as below:

1 Parametrically Mapped Audio Reactive.

   a A single master output feed was taken from the audio stream. The amplitude of the audio was parametrically mapped to shape size and shape position. A large amplitude sound, such as a percussive transient, would cause the shapes to move and/or rapidly expand and contract in size. The intensity of these effects could be adjusted between no effect to large deviations.

2 Interactive HCI. A MIDI controller provided the ability to control the parameters below (brackets indicate type of physical controller)

   a X and Y spacing of shapes (rotary dial)
   b Primitive shape size (rotary dial)
   c Shape distribution around a 3D sphere (rotary dial)
   d Colour cycling rate, active when object colour made visible by controller 'l' (see below) (rotary dial)
   e Full image strobing effect (rotary dial)
   f Selection of black and white or full colour noise effect (push button)

g Primitive shape type, circle, torus and grid (rotary dial)
h Number of shape repetitions (fader)
i Noise amplitude—shape distortion (fader)
j Shape size audio reactivity (fader)
k Shape position audio reactivity (fader)
l Colour/monochrome crossfade (fader)
m Shape line thickness (fader)
n Master level fader (fader).

The audio reactive elements described in '1' above still have a degree of control whereby the produced effect can be bypassed or exaggerated by a setting on the control surface. Each of the HCI elements in '2' was assigned to either a unique fader, rotary dial or button on the control surface. These assignments are listed in brackets. Faders were used on parameters that were used most frequently and where most travel and resolution were required. Rotary dials were used less frequently. A switch was used for the parameter that had only 2 states, on and off.

Conversely to the interactive qualities of the visual instrument was the capability for it to permit a degree of *free running*. In this sense free running is the ability of the DVI to sustain visual continuity and flow without the necessity for continual interaction. It is not necessary to reflect every musical quality or gesture with a corresponding visual response. At various points in the performance, it may be suitable to let the visuals continue in their current state, perhaps to let the music undergo independent development. To achieve this the MIDI controller could be set to specific settings whereby there was sufficient visual variation to require no further input. This entailed adjusting audio-reactive elements, colour cycling and noise distortion to cause suitable free-running visual continuity.

Biphase Compositional Form

Compositionally, Biphase has two key sections, which mirror the musical composition. After beginning with two distinct circular forms, hinting at the title, the images become more textural in nature, and these qualities persist through the first section of the piece. Textures are used here as they reflect similar qualities in the gradually evolving musical drones. After a brief pause the second section commences with clearly visible figures which respond gesturally to the more rhythmic nature of the music. In terms of interaction, the first two-thirds of the piece required only very gradual and subtle changes. During the final third of the performance the music becomes more intense and dynamic, meaning increased interaction was necessary. The development of the piece and audio-visual relationships which were considered are described below. A full render of the piece can be viewed on YouTube here: /www.youtube.com/watch?v=HXwikHiP4Hg (Payling and Dooley 2022b).

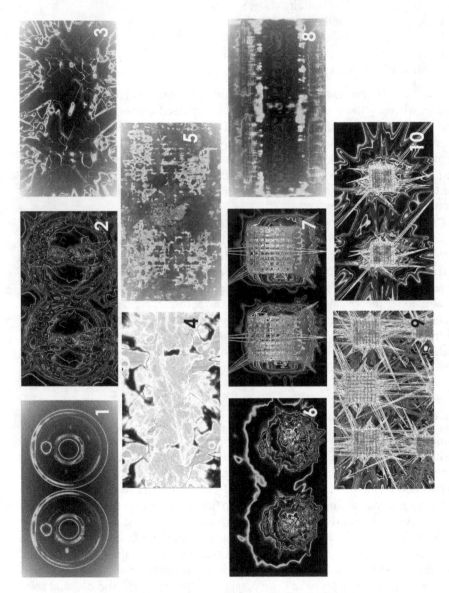

Figure 3.2 Biphase Visual Music Performance Timeline. Screenshots from Key Structural Phases

Audio-visual Development of Biphase with Timestamps

Screenshots accompanying the below descriptions can be viewed in Figure 3.2.

1 00:00–2:20. Introduction of dual circular forms which expand as the drone texture intensifies. Two repeated shapes are the central motif to align with the title '*Biphase*'.

2 2:20–3:25. The circular forms begin to distort to reflect the developing glitchy rhythms and additional tonal textures.

3 3:25–4:00. The circular forms become less coherent, and noisy, displaced lines follow the flow of the textural audio drone and glitches. The circles merge into each other.

4 4:00–5:12. Dominant, pitched musical tones introduce brighter washes of blurred light and colours. An expansive colour space and shifting light replaces the distorted lines.

5 5:12–5:54. Flashes of alternate, square-shaped, noisy forms are quickly introduced and removed to pre-empt the change of emphasis which occurs during the second part at 6:40.

6 5:54–6:40. Dual circular noisy forms are reintroduced and they move randomly around their anchor point to emphasise the glitchy rhythms.

7 6:40–8:28. The mood of the music changes. It becomes more sparse and percussive sounds take the foreground. The shape converts to a distorted square box with fragmented and spiky noise lines. The squares move rapidly in sync with the rhythm. The square shapes continue to morph with some subtle colour transitions.

8 8:28–9:56. Audio noise and high-pitched blips signify changes in visual colour and shape. The percussive musical rhythm continues to develop and increase in complexity.

9 9:56–10:47. The shape motion intensifies; it accentuates the flow of the percussive rhythms. Additional repeated shapes are introduced to further reinforce the rhythmic-visual relationship.

10 10:47–11:41. The sonic intensity begins to recede as sound layers are removed. The shapes continue to move in unison with the rhythm and the visual takes priority over the musical. The shapes gradually reduce their motion as the composition draws to a close.

Notes

1 Lights can take on an informative role, for example in illuminated signage.
2 Not to be confused with the open *sound* control (OSC) standard which is a networking protocol designed for networking musical devices which can be considered an enhancement to the MIDI protocol.

References

Bernard, David. 2006. 'Visual Wallpaper'. VJ Theory. www.ephemeral-expanded.net/vjthe ory/web_texts/text_bernard.htm.

Bongers, Bert. 2000. 'Physical Interfaces in the Electronic Arts: Interaction Theory and Interfacing Techniques for Real-Time Performance'. In *Trends in Gestural Control of Music*, edited by Marc Battler and Marcelo Wanderly. Paris: IRCAM.

Carvalho, Ana and Cornelia Lund. 2015. *The Audiovisual Breakthrough*. Germany: Collin & Maierski Print.

Ciciliani, Marko, Barbara Lüneburg and Andreas Pirchner. 2021. *Ludified: Band 1: Artistic Research in Audiovisual Composition, Performance & Perception/Band 2: Game Elements in Marko Ciciliani's Audiovisual Works*. Berlin: The Green Box.

Conrad, Tony, dir. 1966. *The Flicker*. Short.

Cooper, Max, dir. 2017. *The Science Behind Emergence AV Lecture*. In Science Film Festival 2017. Nijmegen, Netherlands. www.youtube.com/watch?v=VFjIk_CnRUM.

Csikszentmihalyi, Mihaly. 2008. *Flow: The Psychology of Optimal Experience*. New York: Ingram International.

Csikszentmihalyi, Mihaly, Monica N. Montijo and Angela R. Mouton. 2018. 'Flow Theory: Optimizing Elite Performance in the Creative Realm'. In *APA Handbook of Giftedness and Talent*, edited by Steven I. Pfeiffer, Elizabeth Shaunessy-Dedrick and Megan Foley, pp. 215–29. APA Handbooks in Psychology®. Washington, DC: American Psychological Association. https://doi.org/10.1037/0000038-014.

Doucet, Jean-Emmanuel. n.d. 'Open Stage Control'. Open Stage Control. Accessed 10 August 2022. https://openstagecontrol.ammd.net/.

Estibeiro, Marc and Dave Payling. 2020. 'Heatwork—Live Audio Visuals and the Potteries'. Performance presented at the SABRE Festival, Zurich University of the Arts (ZHdK), Pfingstweidstrasse 96, 8005 Zürich, 15 February. www.sabre-mt.com/events.

Foakes, Kevin. 2022. *Wheels of Light: Designs for British Light Shows 1970–1990*. Four Corners Books. https://coloursmayvary.com/products/wheels-of-light-designs-for-british-light-shows-1970-1990-kevin-foakes.

formuls. 2018. *Beforethetitrationofpanacea, by Formuls*. Digital Release. Birmingham: Bandcamp. https://formuls.bandcamp.com/album/beforethetitrationofpanacea.

———. 2019. *Afterthedeathofego*. Digital Release. Birmingham. https://formuls.bandcamp.com/album/afterthedeathofego.

Gibson, Steve. 2022. 'Liquid Visuals: Late Modernism and Analogue Live Visuals (1950–1985)'. In *Live Visuals*, edited by Steve Gibson, Stefan Arisona, Donna Leishman and Atau Tanaka, pp. 67–88. London: Routledge.

Gibson, Steve and Stefan Arisona. 2022. 'VJing, Live Audio-Visuals and Live Cinema'. In *Live Visuals*, edited by Steve Gibson, Stefan Arisona, Donna Leishman and Atau Tanaka. London: Routledge.

Harris, Louise. 2016. 'Audiovisual Coherence and Physical Presence: I Am There, Therefore I Am [?]'. *EContact!* 18 (2). http://econtact.ca/18_2/harris_audiovisualcoherence.html.

Hermes, Kirsten. 2021. *Performing Electronic Music Live*. 1st edition. New York: Focal Press.

Joaquim, Vitor and Álvaro Barbosa. 2013. 'Are Luminous Devices Helping Musicians to Produce Better Aural Results, or Just Helping Audiences Not to Get Bored?' In *Conference on Computation, Communication, Aesthetics and X*, 1: 89–105. Bergamo, Italy: xCoAx.

Kester, Walt. 1999. 'Introduction'. In *Practical Design Techniques for Sensor Signal Conditioning*. Wilmington, MA: Analog Devices.

Manchine, dir. 2016. *Alva Noto—Live in Copenhagen 2016—FULL CONCERT*. Copenhagen. www.youtube.com/watch?v=Y_-hKHbO-GE.

Miranda, Eduardo Reck and Marcelo M. Wanderley. 2006. *New Digital Musical Instruments: Control and Interactions Beyond the Keyboard*. Middleton, WI: A-R Editions, Inc.

Moolhuijsen, Martin. 2021. 'Seeing Time Through Rhythm: An Audiovisual Study of Flicker Films'. *Sonic Scope: New Approaches to Audiovisual Culture*, no. 3 (October). https://doi.org/10.21428/66f840a4.cc0b43a4.

Orgeman, Keely. 2017. *Lumia*. New Haven, CT: Yale University Press. https://yalebooks.co.uk/page/detail/?k=9780300215182.

Payling, Dave. n.d. 'NoiseFloor UK'. NoiseFloor UK. Accessed 21 November 2018. https://noisefloor.org.uk/.

Payling, Dave and James Dooley. 2022a. 'Biphase'. In *International Computer Music Conference*. Limerick, Ireland: International Computer Music Association. http://eprints.staffs.ac.uk/7357/.

Payling, Dave and James Dooley, dirs. 2022b. *Biphase Live @ ICMC 2022*. Live Performance. Limerick. www.youtube.com/watch?v=HXwikHiP4Hg.

Sá, Adriana and Atau Tanaka. 2022. 'A Parametric Model for Audio-Visual Instrument Design, Composition and Performance'. In *Live Visuals. History, Theory, Practice*, edited by Steve Gibson, Stefan Arisona, Donna Leishman and Atau Tanaka 237–66. London: Routledge.

Sanden, Paul. 2013. *Liveness in Modern Music: Musicians, Technology, and the Perception of Performance*. Electronic book. Routledge Research in Music 5. New York: Routledge.

Schechner, Richard. 2003. *Performance Theory*. 1st edition. New York: Routledge.

Schechner, Richard and Sarah Lucie. 2020. *Performance Studies: An Introduction*. 4th edition. London: Routledge.

Sonobus. n.d. 'SonoBus'. Accessed 10 August 2022. www.sonobus.net/.

Spahn, Claudia, Franziska Krampe and Manfred Nusseck. 2021. 'Live Music Performance: The Relationship Between Flow and Music Performance Anxiety'. *Frontiers in Psychology* 12 (November): 725569. https://doi.org/10.3389/fpsyg.2021.725569.

Steinman, Robert M, Zygmunt Pizlo and Filip J Pizlo. 2000. 'Phi Is Not Beta, and Why Wertheimer's Discovery Launched the Gestalt Revolution'. *Vision Research* 40 (17): 2257–64. https://doi.org/10.1016/S0042-6989(00)00086-9.

Steve Pavlovsky. n.d. 'A Brief History of Psychedelic Light Shows'. Tumblr. Accessed 4 November 2022. https://liquidlightlab.tumblr.com/.

Trevor, National, dir. 2018. *Brains Need Bodies Too*. Digital Performance. www.youtube.com/watch?v=tMOfAUw9tr0.

Tsay, Chia-Jung. 2013. 'Sight over Sound in the Judgment of Music Performance'. *Proceedings of the National Academy of Sciences* 110 (36): 14580–85. https://doi.org/10.1073/pnas.1221454110.

URBANSCREEN, dir. 2012. *SYDNEY OPERA HOUSE | Facade Projection | Extended Version*. https://vimeo.com/45835808.

Vuoskoski, Jonna K., Marc R. Thompson, Charles Spence and Eric F. Clarke. 2016. 'Interaction of Sight and Sound in the Perception and Experience of Musical Performance'. *Music Perception* 33 (4): 457–71. https://doi.org/10.1525/mp.2016.33.4.457.

4   Compose

Electronic Visual Music

Composing Electronic Visual Music

Achieving the ideal of an engaging and cohesive visual music composition is a challenge in both technical and artistic terms. This chapter will detail several creative strategies to help the reader rise to this challenge. Compositional processes are discussed in relation to fixed media; this means visual music videos that are to be presented through screening, as opposed to the real-time manipulation and performance of media as discussed in the previous chapter. As with all chapters in this book, it is written from the perspective of a musician and coloured by my experiences of teaching these techniques to music students. It will hopefully provide insight for musicians and video artists alike.

Although multidisciplinary practice is becoming more common, many artists still associate themselves closely with a single discipline, such as music or video art. It can be daunting to attempt to extend these capabilities and adapt the years of experience and intuition acquired in one's familiar territory into a second discipline. Fortunately, many aspects of compositional expertise translate intuitively between disciplines and can help streamline the process of mixed media composition. When facilitating composition through digital platforms there is a surprising level of overlap between music and video workflows, albeit with peculiarities related to discipline-specific terminology and toolsets. Even so, composing for multiple media presents a degree of challenge greater than working in one medium alone. The music composer shapes a composition with singular attention to make each musical element blend with the others in a cohesive manner. Visual composition can also be a similarly involved process, visual aesthetics require time to develop and the formulation of raw materials into satisfactory compositions takes intuition and practice. Bringing both media together adds another dimension that is greater than the sum of the parts. Novel and surprising results reveal themselves causing the artist to re-evaluate their original expectations. This is the phenomenon whereby 'one perception influences the other and transforms it. We never see the same thing when we also hear; we don't hear the same thing when we see as well' (Chion 1994, XXVi). This is a subjective response that the composer continuously deals with, and it is these unexpected emergent synergies which keep things challenging and rewarding both for the artist and the viewer.

DOI: 10.4324/b23058-5

Compositional Approaches in Visual Music

Many compositions originate with a concept; an idea, theme or technology is chosen on which to base the finished work. Although this is not essential, a concept gives a project direction, something which can be continually referred to when deciding if things are moving in the desired direction. The choice of concept is at the composer's discretion and can come from many sources as evidenced by the interviews in Chapter 2. The initial ideas can be developed into a plan with as little or as much detail as the composer prefers. Traditionally in music, creating a composition would involve music notation or a graphic score; in video production a storyboard might be used, and these can still be used. Personally, I usually have an idea and loose structure for the piece, which might have a few notes, but is largely an abstract concept I mentally refer to while working. One overarching goal is to create sound and vision with an equitable relationship. Visual music places importance on this aspect of the resultant work. Sometimes in a piece, however, one medium might be more dominant than the other due to the choice of theme, the composer's intent or the subjective reception by the audience. In practical terms it may also arise from the approach taken during composition.

During composition one medium may be completed before commencing work on the other. This can be referred to as visual inception or music inception. The term 'inception' is used here to highlight that, from the concept, the composition is commenced in the chosen medium. Visual inception is perhaps the most common and is used frequently in cinema where the film can be finished entirely before being handed over to the sound department to complete the soundtrack. This can be a fruitful exercise for the musician. Once fluency is developed interpreting visual scenes, the process of composing sound to picture becomes intuitive. Freed from the uncertainty, which is the blank music project, the composer has a concrete set of images and narratives from which to be inspired. This still applies if the artist creates their own video materials. Visuals can be a rich source of inspiration from which to compose. The video can also carry the flow of the piece, so it isn't necessary for the music to be as richly detailed as might be necessary when composing music alone.

The music first approach is possibly the route a musician might take. Being more comfortable creating a soundtrack, there can be a tendency to complete the familiar part before adding visual imagery to complement it. Animating to an existing soundtrack can be an effective approach, as visual events can easily be synchronised to musical events. This is the approach used by Fischinger and other animators who, for some pieces, would choose a score from which to be inspired. A more difficult task is compiling real-world video footage to a soundtrack and maintaining visual continuity and cohesion. There may be gestures and transitions in the sonic domain that require an equivalent response in the visual domain. Attempting to convey this through video footage requires careful selection of appropriate scenes or employment of the montage techniques described below.

Apart from single medium inception there is a further method, the bi-directional, or concurrent approach.[1] This technique, favoured by Harris (see interview in

Chapter 2), is a more fluid and iterative approach where each medium is partially completed before switching to work on a subsection of the other medium. This can be more time-consuming as it is necessary to evaluate each stage of the composition as it is developed. It will usually involve some degree of music or visual production to get things moving. One might, for example, compose a section of music to a partially completed video. The results of this are observed and the composer may then decide to modify some elements of the visuals to improve overall coherence. This coherence and sound-image synergy can be evaluated with reference to the original concept or the relationships emerging between the media themselves. Such relationships can include those related to synchronisation, colour and timbre and others as discussed in Chapter 1. Once the composer is satisfied with the results thus far it might then be necessary to further adjust either medium so it integrates even more closely with the other. This process continues in a cyclical manner until the decision is made to move to conclusion, or onto the next step. This 'non-committal' aspect is the one that can create more work and extend the whole process. It can however result in greater cohesion between sound and image as they are constantly being evaluated and attuned to bind more closely; the sound and image relationships are being created at a finer granular precision. Successful results can, however, still be achieved by completing a single medium as discussed previously; having one element in its finished state imposes a type of creative constraint that can be very productive. Ultimately there is no right or wrong approach and the composer may employ a different method each time or a combination of each.

Material Transference and Compositional Thinking

Two concepts related to the interplay between video and music are relevant here, they are *material transference* and *compositional thinking* (Hyde 2012). These are methods for applying musical constructs to video and vice versa and are useful considerations when using any of the above approaches to composition. *Material Transference* is an 'area of practice that involves a process of material transference from music (or sound) to the ocular domain . . . This approach can be broadly described as the expression of tone in both aural and ocular domains, where equivalence is sought between the frequency domain in sound and light (colour)' (Hyde 2012, 170). This definition can be expanded to encompass the montage techniques discussed later and all bi-directional, direct or indirect influences between sound and image. Material transference would be achieved, for example, through a parametric mapping process to cause characteristics such as sound amplitude, to modify the size of a shape. These types of relationships between sound and image can also be considered and developed while in production. *Compositional thinking* on the other hand is a practice which 'involves the application of more abstract ideas and principles derived from music to ocular media' (Hyde 2012, 171). Music by nature is abstract; it has no reference to representational imagery in the same way that painting might. It has abstract principles related to structural form, dynamics and temporal development and so on, some of which can be applied to the temporal organisation of visual composition. The composer may, for example, decide to impose a binary AB structure onto the video composition with two distinct but related sections. Compositional thinking can be thought of as a top-down approach to composition, whereby higher-level structures are imposed to determine the arrangement of media materials. They are useful at the planning stage. Material transference is more closely aligned with a bottom-up approach, working with the materials themselves to develop the composition. As will be discussed later, material transference also aligns with code-based generative techniques whereas compositional thinking fits more closely with composer rendition.

Audiovisual Landscapes

Audiovisual landscapes are another consideration one might consider when composing. This is a technique borrowed from music (Field 2000) which can be applied to visual music composition. One challenge in music composition is to create a sense that the sounds and structure have a shared musical purpose. Although sounds can be combined easily, and even randomly, this may not necessarily produce a piece of music with coherent thematic qualities. Creating coherence within a soundtrack is as much a mix problem as it is a compositional one. Just one element poorly levelled, spectrally imbalanced or otherwise out of place can undermine the credibility of the whole piece. In landscape terms, combining sounds recorded in different environments can result in a confusing juxtaposition of different environments. A recording of speech captured in a reverberant indoor room may not mix convincingly with an outdoor ambience, for example. One way of avoiding this is to record all sounds in the same environment or in an acoustically dry environment and to add spatial effects afterwards. This environmental coherence can be considered as the development of a *'sonic-landscape'* (Field 2000, 44). A sonic-landscape is created by ensuring the overall aural product results in a cohesive environmental space, whether this be real or not. It is not relevant that this was the intention of the composer at the outset, these landscape morphologies permit retrospective analysis and can be used to inform future compositions. Field's sonic-landscape morphologies consist of four types:

1. 'hyper-real
2. real
3. virtual
4. non-real'.

Hyper-real landscapes contain material that is recognised as real but has been treated in some way to exaggerate or enhance the sound to make it appear 'more than real' (Field 2000, 45). This could be created by field recordings that are filtered, time-condensed or processed in some other way to emphasise certain properties while maintaining a semblance of reality and integrity of the space. The opening minute of the soundtrack of Adkins' *With Love from an Invader* (Adkins 2021) illustrates this point, with field recordings that take you to Shedden Clough at the outskirts of Burnley UK, but subtle fades and processing augment the original realism. *Real* landscapes on the other hand use non-destructive audio processing and simply reproduce the recording without modification. There are many examples from Bernie Krause's natural soundscape recordings, such as *Autumn Day in Yellowstone* (2007), which are real landscapes and place the listener in different locations around the world. *Virtual* landscapes are created by simulation. This can involve reverberant effects and treatment of recordings which create unusual effects and spaces. *Microplastics* (Blackburn 2021) takes sounds of various household plastic objects and transforms and places them in their own virtual world. Although real sounds are captured in real environments, the constructed landscape bears no resemblance to the original place of recording. Finally, heavily processed or synthesised sound can create a *non-real* landscape that 'does not contain any real world gestures or naturally occurring sounds' (Field 2000, 47). Any purely synthesised piece of music falls within this category. *Dancing in the Ether* (Kuehn 2020), for example, is constructed entirely from synthesised sounds and a non-real three-dimensional space is explored through various permutations of the sound materials. Although Field describes these as separate landscape morphologies, they can be combined to create different effects. Creating the impression of a real-world recording moving through a virtual acoustic space, for example, can create an intersection between a real environment and a virtual one but, as explained, care has to be taken to make this work in practice.

Applying Field's morphologies to visual music compositions, *audiovisual landscapes* can be developed, providing another means of analysing and creating compositions, this time with regard to environmental coherence. Just as sonic landscapes inhabit a particular environment, visual music pieces can be said to inhabit an *audiovisual landscape*. When combing media in visual music, this presents the composer with the additional opportunity to combine music from one landscape with visuals from an alternate landscape, potentially creating useful results. To illustrate audiovisual landscapes, example compositions are discussed. A *hyper-real* landscape is present in Monty Adkins and Yan Wang Preston's audiovisual version of *With Love. From an Invader* (2021). The *hyper-real* environment is created through a time-lapse recording, taken over a year, of a single rhododendron tree accompanied by a soundscape of audio recordings collected from the same area. The result is a time-condensed experience of a whole year of natural decay and regeneration recreated and enhanced. A *real* landscape is evident in Cope and Howle's *Flags* (2011). There are some untreated field recordings in the soundtrack, but heavily treated sound recordings are also used. The video consists of just two untreated shots of Buddhist prayer flags, strung across a mountaintop in Tibet. The juxtaposition of alternate morphologies in soundtrack and video results in the slowing of time and places the visual Tibetan environment within a virtual sound space. *Neshama* (Oliveira 2018) is an example of a *virtual* composition. While there is a recognisable human figure in the video, the images are treated to abstraction, creating a unique virtual space. As in *Flags*, the soundtrack to *Neshamah* is an abstract electroacoustic piece, which further enhances the surrogacy from a recognisable real or hyper-real environment. *Non-real* visual music compositions are possibly the most prevalent type in visual music and are especially evident in those created by abstract animation. Battey's *Autarkeia Aggregatum* (2005a) is created both visually and sonically by synthetic means. The resultant landscape has no recognisably real components but it creates its own rich visuo-spatial and sonic textures.

Just as with music composition, there could be a credibility issue created by the poor combination or treatment of individual visual elements. If one element does not fit in the emerging landscape, due to inadequately composited or treated images, it can detract from the integrity of the environment. As an example, imagine selectively colouring an object differently to the rest of the other subjects in an image. This will immediately draw attention to that object and change the perception of the whole scene. Does it still belong here, why does it appear differently and how did it get there? This may of course be the intention and can be used as a creative effect.

Field goes on to discuss transcontextuality whereby landscape morphologics are themselves transformed during the piece. Transitioning between landscapes is enabled by transcontextual agents and can transport the viewer from one space to another. In *Threshing in the Palace of Light* (Piché 2017) there is an intriguing layering of sonic and visual materials. On one level there are abstract evolving geometric visual forms which are expansive and immersive. This is a non-real landscape with cohesive integrity that invites the viewer into its world. Superimposed on this are real actors and glitchy film clips. It is as if the viewer is glancing through the non-real landscape into memories that occurred in a real landscape. The visual materials themselves are also processed by effects which simulate film shutter glitches and disrupt the immersive textural landscape. When these occur, they act as a transcontextual agent causing immersion in the non-real landscape to be lost as we become aware of the technology, or in this case a simulation of the technology. The soundtrack is similarly uncertain, with spoken words blended with steady textural drones. Again, the drones are immersive and non-real, matching the abstract visuals, but the words are very real and from another space. When present, they take the listener away from the non-real ambience into a remembered space where the words were originally spoken. The whole piece gives the impression of being in a dreamscape with lost vocal fragments occasionally entering the conscious.

Gesture and Texture in Visual Music

Another musical strategy that can be used when composing visual music is that associated with gesture and texture. This is a concept first explored in my PhD thesis (2014) and more recently by Knight-Hill (2020). Smalley discussed gesture and texture in relation to electroacoustic music composition (1986, 80–84) explaining:

> Gesture is concerned with action directed away from a previous goal or towards a new goal; it is concerned with the application of energy and its consequences; it is synonymous with intervention, growth and progress, and is married to causality.

> (1986, 82)

Gesture is a movement and development in the musical materials and would commonly be experienced over a relatively short duration. It could be a transient, a glissando or other sound that can be identified as transitioning between states. Acoustic instrumental performance is by nature very gestural; consider the actions of a percussionist and how their physical gestures translate into sonic gestures.

> Texture, on the other hand, is concerned with internal behaviour patterning, energy directed inwards or reinjected, self-propagating; once instigated it is seemingly left to its own devices; instead of being provoked to act it merely continues behaving.
>
> (Smalley 1986, 82)

Textural sound in this context is more static and evolves gradually over time; obvious gestural movements are minimised. Musically, I often refer to the sound created by gong instruments, especially the large symphonic gongs in the region of three feet in diameter. Their sound is initiated by a percussive type of gesture but once struck, the sound continues to resonate long after the stimulus is removed. Texture is also a description that could be applied to long, extended pad-type sounds and drones, generated either acoustically or electronically, that would be present in ambient soundscapes.

Applying this terminology to the visual medium reveals some similarities best explored again through examples. McGloughlin's animation to Max Cooper's *Symmetry* (Cooper, McGloughlin and Hodge 2017) is very gestural in nature. The repeated circles begin to move quite slowly but soon respond more quickly to the sonic events and pace of the music.[2] Synchronisation between edits and variation in their positioning relative to the meter create rhythmic interplay between sound and image. Now turning attention to textural qualities, these can be seen in *Amorphisms* (Miller and Young Ha 2008). This is a piece that uses a delicate, relatively consistent soundtrack to accompany a visual animation constructed from an expanse of interchanging colours. Coupled with the textural sound and image there is no deliberate attempt at synchronisation, as the flowing colour spaces allow little opportunity for this. There are occasional flourishes in both sound and video, but the focus is a sustained consistent atmosphere in music and colour. The soundtrack could be offset in time from the video and the complementary nature of the materials would still be evident. Textural sounds therefore make suitable companions to more abstract colour spaces. In this context, space should be interpreted as the rectangular real estate of the screen dimensions. Conversely, it is possible to mix textural qualities in one medium with gestural qualities in the other. In *A Fleeting Life* (Cooper and McGloughlin 2020) visual gestures are created by rapid edits between different film shots. This is coupled with a textural harmonic drone. The juxtaposition of thoughtful textural music with rapidly moving imagery creates an emotional resonance with the composition's theme.

Texture from Gesture

One technique of creating textural sound is to begin with gestural sources and subject them to a process of time stretching.

> The more gesture is stretched in time the more the ear is drawn to expect textural focus. The balance tips in texture's favour as the ear is distanced from memories of causality, and protected from desires of resolution as it turns inwards to contemplate textural criteria.
>
> (Smalley 1986, 84)

If a gestural sound recording is time stretched beyond any reasonable limit, the resultant sound tends towards a textural drone. The gesture is extended so far that it becomes unrecognisable as such. Similarly, gestural visuals, such as an animated circle, can be made more textural by zooming in and slowing down the frame rate. By doing so, the inherent gestures will be minimised and the textural colour properties emphasised. These are all tricks that can be explored when working with audio-visual materials to manipulate the horizontal time and the vertical inter-media relationships. Visual and musical textures can also be considered as background materials and gesture as foreground materials, although this does not always need to be the case. Objects which are the subject of the visual narrative in a sequence are often gestural in nature and can be accompanied by gestural sounds. Think of animated objects tightly synchronised to sounds. These elements may be foregrounded in the musical and visual mix to reinforce their dominance of the scene. Textural images are more spatial in quality and are potentially useful for backgrounds to the gestural elements.

Thomas Wilfred's Lumia and Visual Music

Lumia is an artform which uses light as its principal medium and whose factors can inform fixed media compositions. The clavilux, 'a name derived from the Latin, meaning light played by key' (Stein 1971, 4) and invented by Thomas Wilfred, was a device that permitted performances with lumia. Lumia are comprised of three primary factors of light, these are form, colour and motion (Wilfred 1947, 247). Compared to colour organs, which commonly mapped musical pitch to the rainbow hue scale, and the paintings of Kandisnksy, in which he employed the elements of colour and form, the clavilux put light and colour into motion, creating an expressive artform which manipulated light. Wilfred designed and made several lumia, generating clavilux instruments, some of which were self-operating and internally programmed and others that could be performed in real-time. The visual output of the instruments was a light projection that was capable of creating complex visual arrangements:

> A typical composition contains one principal motif with one or more subordinate themes. Once chosen, they vary infinitely in shape, colour, texture, and

intensity. The principles evident in plastic and graphic compositions—unity, harmony, and balance—function kinetically in lumia. When movement is temporarily suspended in a lumia composition, the result is a balanced picture. However, the static picture's ultimate meaning can only be seen in relation to what follows it.

(Stein 1971, 3)

Wilfred described lumia in detail; the primary factors also had four sub-factors each, and he grouped these into a 'graphic equation' (1947, 254). The combination of light's lumia factors, to the left of the equation, produced the artistic potential on the right:

Place this inert potential in a creative artist's hand, supply him with a physical basis—screen, instrument and keyboard—and when his finished composition is performed, the last link in the chain has been forged and you have the eighth fine art, lumia, the art of light, which will open up a new aesthetic realm, as rich in promise as the seven older ones.

(Wilfred 1947, 254)

Wilfred suggested he had laid the foundations for a new artform that would continue to be studied and practised long into the future, but there was limited work subsequently using lumia or the clavilux instruments, until a recent revival of interest by academics and artists. Lumia and Wilfred's discussion of the artform can, however, be used as inspiration when creating visual music compositions and performances. The following techniques were first discussed in eContact!, the online journal for electroacoustic practices (Payling 2017).

Lumia Form in Visual Music

In some artistic disciplines the subjects and forms of the artwork are an important focus, rather than the colours from which they are created. In classical sculpture for example, a bust may be carved, or otherwise formed, from a single material of uniform or near-uniform colour. The sketching artist may use a graphite pencil to quickly draw the shapes of people and other objects. In photography, black and white images are still used and occasionally films, either completely or in part, are made without colour. The subjects or objects in films are often the important focus, possibly because they are something with which the audience can relate or identify. Colour can often be omitted without modifying the visual narrative. Visual music compositions also frequently contain objects or forms, however abstract, whose behaviour is central to the composition. Such objects are rarely static, and their movement is often used to inform the modulation and tempo of the music. Moritz (2004) quoted Fischinger's work plan for constructing *R-1 Formspiel* (Form Play) and some notes for *R-2* which he believed were related, which highlights this point:

Points of Staffs start to dance slowly, rhythmically, and arise gradually up to the middle of the picture. The tempo at first is practically non-existent, and then begins only slowly to become perceptible. Now single pieces grow out above the general line and take the lead in a particular way. Then they destroy the uniform line and attempt to lead an individual, independent life.

(Fischinger 2004, 176)

These notes describe an abstract piece that concentrates on the evolution and movement of variously formed objects. Although very descriptive in terms of movement and pace, the abstract forms, which the objects possess, appear to play an important role in the design of this piece. The work *FLUX* (sisman 2010) begins with a red circle which Complex forms emerge and reshape and are mirrored by a soundtrack which mirrors these transformations. It is difficult to separate the form from the movement to which the circle is being subjected but, as it shifts and reforms, it is reflected by changes in the soundtrack.

Lumia Colour in Visual Music

Can colour exist without form and motion? Possibly the closest we can get to this state visually is a colour field. A colour field is an area of colour that completely engulfs the visual domain and can be described as 'an area of any single homogeneous color extending as far as the observing eye can see at that moment. This translates, in experiential terms, to the experience of seeing nothing but, say, light green' (Sloane 1989, 168). In practical terms, this visual sensation could be achieved by holding a uniform matte surfaced piece of monochromatic coloured card near to the eyes. With uniform diffuse lighting, a constant colour would be perceived. If the card were to be moved away from the eyes, the colour would eventually take on the rectangular form of the card and the experience of pure colour would be lost. This is similarly the case for colours that are rendered onto a computer or other screen. Although the screen may be filled with a single colour, it is constrained and defined by the physical shape of the display. In painting, a canvas filled with a single hue would be very minimal, but some artists have created works that approach this. One such artist is Barnett Newman who helped develop the minimalist painting artform known as colour field painting where large canvases were filled with several layers of a single colour. *Cathedra* by Barnett Newman (1950) used six layers of different blue pigment that were applied to a 2.5 metre high by 5 metre wide canvas with the addition of two vertical 'zips' which separate expanses of blue. The result is a richly nuanced surface and field of colour. This painting is best experienced closely and can create spatial illusion as it lacks a focal anchor point. Connecting with the colour, depth and the observer's interaction and engagement with the canvas are key to experiencing this painting.

In contemporary visual music, colour can be used in relatively pure ways with abstract random forms filling the frame, but a more vivid example of the exclusive use of colour in film is *Color Sequence* by Dwinell Grant (1943). A forerunner to the flicker films of artists including Tony Conrad and Paul Sharits, this silent film switches between coloured frames at varying rates creating rhythmic effects due to the changes in hue and intensity. This approaches one of the purest experiences of colour possible in the film medium. Referring to the previous discussion on *gesture* and *texture*, colour is also a useful factor when constructing textural elements in visual music.

Lumia Motion in Visual Music

Whereas colour associates naturally with *texture*, motion aligns more intuitively with *gesture*. An artistic discipline making use of motion is dance. As with the previous factors of form and colour, motion cannot be isolated completely, a dancer will hold various forms and may be colourfully dressed. In dance, however, the movements of the dancer, often performed in a rhythmic way, are the key focus. Dance routines often also put music into motion through tightly choreographed performances. In film, recorded footage implicitly contains motion. Even if the camera recording a scene remains in a fixed position, there will be some movement of light and shadow over time. If a series of still images were presented as a film or slideshow there would be movement, or light changes, as the images were transitioned. These are, however, quite minimal examples of motion; it is more common in visual music for objects to move through a trajectory, often in a gestural manner. An early animated film exploring temporal design and motion is *Symphonie Diagonale* (Diagonal Symphony) by Viking Eggeling (1924). This piece could also be interpreted as a study in geometric form, but the diagonal movement of abstract objects dictates the pace of the animation. A less overt approach, but nonetheless effective use of motion, is in the film *Heliocentric* by Semiconductor (Jarman and Gerhardt 2010). This piece used time-lapse recordings of the sun passing across various landscapes. The motion of the sun is tracked and accentuated by the music, which is predominantly constructed from a high-pitched ringing tone. As the name, *Heliocentric* suggests, the sun is always at the centre of the screen. The audible ringing tone, which depicts it, is therefore always centrally panned in the stereo image. As objects in the landscape obscure the sun, the ringing tone becomes quieter. When the screen is brightened, as clouds diffuse the sunlight, the sound becomes louder. Although simple in concept, the quality of the video recordings and the attention to detail in the soundtrack combine to make the motion of the sun very apparent, although, in actuality, it is the earth and landscape which is moving.

Combinations of Form, Colour and Motion

A further way of considering the lumia factors is how they pair with each other. The possible combinations are illustrated in Figure 4.1, and Wilfred summarised:

> Form, color and motion are the three basic factors in lumia—as in all visual experience—and form and motion are the two most important. A lumia artist may compose and perform in black and white only, never using color. The use of form and color alone—static composition with projected light—constitutes a less important, but still practical field in lumia. The only two-factor combination that cannot meet the requirements is motion and color,

without form. This is because it violates a basic principle in vision. The human eye must have an anchorage, a point to focus on. If a spectator is facing an absolutely formless and unbroken area of color, his two eyes are unable to perform an orientational triangulation and he will quickly seek a visual anchorage elsewhere, an apex for the distance-measuring triangle that has its base between the pupils of his eyes.

(Wilfred 1947, 252–53)

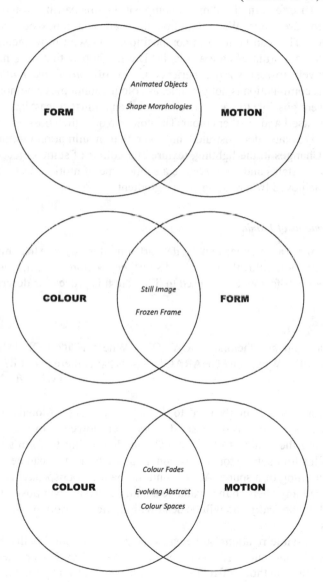

Figure 4.1 Diagrammatic Representation of the Combinations of Lumia Factors
Based on Wilfred (1947)

Form and motion were considered by Wilfred to be the most important factors of lumia. They also pair and occur naturally in film and animation, with moving objects commonly being the focus of abstract compositions. Colour is not essential, monochromatic animations can still hold visual interest as evidenced in *Symphonie Diagonale* (Eggeling 1924). *Colour and form* is the next combination, and can be achieved by the projection of coloured light. This combination would be experienced at music concert light shows and colour organs would have produced a similar effect. In visual music composition, this pairing would equate to a stationary, coloured form. Without motion, the shape and light would remain static on the screen. This can be used in compositions, but with movement, additional compositional opportunities are available. The lumia pair that does not naturally relate intuitively to real-world experience is that of *colour and motion*. Wilfred suggests this combination is not possible, as colour cannot move without holding a form. If it were possible to see a colour move, then it must at least have a boundary line between itself and another colour. The colour would then possess a basic form. McLaren (1978) later demonstrated, however, that in animation motion can relate to change. Changes in the lighting, texture and colour of static objects are movements between states and are therefore a subtle type of motion. In practical terms this can be achieved by a rotating colour gradient.

Sonic Equivalents of Lumia

Although lumia was intended to be a silent artform, it is possible to compare its factors with their sonic equivalents. Wilfred's further division of form, colour and motion into their sub-factors can be used to illustrate this process as described below.

Sound Form

> Form has four sub-factors: LOCATION—Where is it? VOLUME—How big? SHAPE—What is it? CHARACTER—What is there about it?
>
> (Wilfred 1947, 253)

In music, form is commonly used to describe the compositional structure of a piece. Wilfred's definitions of its sub-factors refer, however, to individual lumia. What then are the qualities of individual sounds relating to form's sub-factors? The first of his four sub-factors is location. In audio terms this can be related to the spatial positioning of a sound. Across a multi-channel loudspeaker array, a sound can be panned anywhere in the available sound field. This pan position could also be changed dynamically and will be dealt with in the description of motion and its sub-factors.

Another possible relationship between location and sound is the one between physical height and pitch. This concept was addressed by Roger Shepard (2001) when examining methods of representing pitch, height and something he labels chroma.[3] In this discussion, chroma is the pitch of a note, independent of its octave.

The note 'C' is a chromatic value, but there is a middle C, the C an octave above this and so on. Shepard went on to say:

> in the world, height and pitch are almost always linked, and things that ascend in height usually ascend in pitch.
>
> (Shepard 2001, 160)

Although a commonly used concept, objects that move in the vertical plane are not necessarily represented by a changing pitch value, but it could be a consideration during composition. Volume, form's second sub-factor, can be associated with a sound's amplitude. This has a basis in physical phenomena:

> In the environment, when a sound approaches the listener its spectral and dynamic intensity increase at a rate proportional to perceived velocity. Moreover, the increase in spectral intensity permits the revelation of internal spectral detail as a function of spatial proximity.
>
> (Smalley 1986, 68)

As an example, an approaching sound could be one produced by a car. As it approaches, it is perceived visually as becoming physically larger. The sound of the engine becomes louder to accompany this. More subtle details and higher frequency content of the car's motion also become discernible as it gets closer to the listener. The two remaining sub-factors, shape and character, are suggestive of timbral qualities. Timbre applies more naturally to colour rather than form and will be examined below. One other relationship that could apply, however, is to the sound's envelope. Percussive and bowed sounds, for example, have very different characters and envelope shapes. Percussion usually has a sharp onset transient and a slightly longer decay, whereas a bowed instrument can possess an equally longer onset and decay. The shape of a sound in this context is therefore related to how its intensity evolves over time.

Sound Colour

> HUE—What color is it—red, green, blue? CHROMA—How much gray has been mixed with the pure hue? VALUE—How much white in that gray? INTENSITY—How strong is the light it sheds?
>
> (Wilfred 1947, 253)

Colour has several relationships to sound and Wilfred's sub-factors of hue, chroma and value can be equated to the hue, saturation and lightness (HSV) colour system. This system was used to develop a sound selection system for sonification applications (Barrass 1997, 127–37) and is one possible method of mapping between colour and sound in a more precise way. In Barrass' experiment, hues were mapped to different timbres, saturation to their pitch and value (lightness) to how much high-frequency content they contained. Also refer back to the historical context chapter for more relationships.

The final sub-factor, intensity, could once again refer to the sound's amplitude. This appears to be a duplication of the volume sub-factor of form but does not confuse the issue as a large, intensely coloured object could also be represented with a loud sound.

Aside from Wilfred's definition, colours can be described as possessing spatial qualities. They fill the space between painted lines and object forms and could therefore be portrayed in music with sounds that naturally fill the aural space. This can be accomplished with sounds of longer duration that would fill the temporal space. Extended sounds of this nature could be more textural, with gradual envelope characteristics and a focus on timbral transformation; what one might commonly refer to as a pad sound.

Sound Motion

> ORBIT—Where is it going? TEMPO—How fast? Speeding up? Slowing down? RHYTHM—Does it repeat anything? FIELD—Is it constantly visible, or does any part of its orbit carry it beyond the range of vision?
>
> (Wilfred 1947, 253)

The third lumia component, motion, also has four sub-factors. Orbit and tempo can be associated with the spatial positioning of sounds in a similar way to the location sub-factor of form, but in this case motion is applied. If an object moved across the screen from left to right this could be accompanied by a sound that also panned left to right. The faster the object moves, the more rapid this audio pan would be. Composers often use this effect and several examples can be found in *End Transmission* (Hyde 2009). Rhythmic movement of on-screen objects could also be represented by a similar panning of the sound, or by a percussive or melodic musical rhythm. The 'field' sub-factor is less intuitively linked to sounds but could be said to apply to one that becomes silent. An object 'beyond the range of hearing', in this case, would not be heard. Alternatively, if a sound was tracking an object that moved off screen it could actually appear to also move off screen. This is a phenomenon referred to as 'spatial magnetization' (Chion 2009) and is perceived even if there is only a single speaker producing all the sound.

> If the sound that comes from the fixed speaker is attributed to an on-screen character, and if we see him or her move to the right, we are going to hear the sound move to the right; if the character exists offscreen, we hear the sound as outside the screen too. The phenomenon of spatial magnetization, whereby our attribution of a sound's location depends on what we see of the real or supposed source, can be observed on countless occasions every day.
>
> (Chion 2009, 248)

Lumia Summary

The preceding analysis of visual music in relation to lumia factors and their sonic equivalents can be used in the development of compositions and performances. It is not necessary or useful to implement all the audio-visual analogies in a literal sense, but they can help inform some compositional detail. Form and motion combine quite naturally and perhaps this is why Wilfred considered these as the most important lumia factors. Motion is frequently used in animation whereby objects move in space but it can also relate to how forms are transformed to produce a constantly changing visual experience. Changes in light intensity, often in a strobing manner, also suggest movement. Dynamic movement, transitions and edits in video footage encourage tight synchronisation between sound and image. This may not be possible without certain problems; it is difficult for sound to follow a straight video cut from one scene to another. Translating this literally would mean an abrupt switch between two different sound recordings. It may be more appropriate to soften transitions by using crossfades, or through the addition of reverb and delay effects that extended the sound for a short time after the edits. Although the audio will not change as immediately as the image, the illusion of a rapid audio-visual switch can be maintained.

Isolating colour completely in visual music composition is often impractical. It is more commonly used to colourise objects or backgrounds. When using colour, focus can be shifted towards the textural qualities and away from gesture. A consequence of minimising form and motion in this way can result in the visual composition having limited punctuation and transients, which would translate to reduced dramatic musical gestures that might otherwise accompany animated forms. From a musical perspective, sounds with greater spatial characteristics are washes of sound, sonic air brushes or what in synthesis terms are commonly referred to as pad sounds. Using this type of sound, its timbre, pervasiveness and gradual transformation are more important than its pitch articulation and phrasing. Apart from synthesiser pads, transformation and time stretching of found sound recordings results in timbres having a textural quality. Outdoor field recordings also inherently contain their own spatial qualities acquired from the environment in which they were recorded.

Lumia factors can influence the production of visual music fixed media. Slowly changing colours pair naturally with evenly paced textural sounds, and form and motion create a more dynamic and synchronous audio-visual experience. Lumia's form and motion factors apply most naturally to visual music composition. The sub-factors relating to position, movement, shape and volume could be portrayed easily in both sound and image. The gestural behaviours of moving objects and edits in video also suggested similar qualities in music. Composing to the dynamically changing video is therefore a relatively intuitive task. Conversely, composing music to textured colour can prove to be more of a challenge. With limited visual change and movement, sounds may need to be sustained for longer periods. This will result in little visual or auditory punctuation, creating stasis and repetition in the visual music experience. Even though gesture is minimised, it does present opportunities to work more intently on gradual sound transformations.

Montage Techniques

Montage is a technique (Eisenstein 1949, 72–83) relating to the ways film footage can be edited and arranged to achieve various results. It can be a useful technique when creating visuals but also as a means for *material transference* and *compositional thinking*. Eisenstein's five types of montage are:

1. Metric Montage
2. Rhythmic Montage
3. Tonal Montage
4. Overtonal Montage
5. Intellectual Montage

These montage techniques were named by a film creator and theorist, so may not have the connotations a musician might expect. *Metric montage* uses musical meter to determine the positioning of the video edit. Scene durations are determined relative to measures in the musical score ensuring visual flow is synchronised to the musical meter. A simple rule could be to change the camera angle of a shot at the end of every bar for example. In more abstract compositions any aspect of the video footage or animation could be designed to coincide with temporal events in the music. This type of montage could be utilised when visual music composition begins with a completed soundtrack. Try editing film footage at the end of each bar of music and you should see some instantly rewarding results. If the music has a discernible time signature it can be relatively straightforward to calculate and execute video cuts at various measures within the music. In compositions with no rigid time signature this approach, however, could be less effective. *Rhythmic montage* refers here to the rhythmic flow of the images, not the rhythm of the music. The duration of each edit is not synchronised with the metrics of the musical score; the edits can be of varying duration and are designed to create rhythmic continuity and flow in a visual sense. The editor will cut the footage based on their experience and instinct while considering elements such as emotion, story, rhythm, eye-trace, two-dimensional and three-dimensional spatial continuity (Murch 2001, 18). For visual music composition these decisions should also take into account the qualities and flow of the soundtrack.

Whereas metric montage uses music as the primary influence on the edit, rhythmic montage and the remaining montage techniques are designed to create various results in visual flow and broader meaning derived from the images. In *tonal montage*, flow is used in a wider sense. This type of montage is based on the characteristic emotions and general tone of the piece. The edits are undertaken to maximise the emotional tone of the images. *Overtonal montage* uses a combination of the preceding techniques to elicit the desired reaction from the audience. In visual music this gives the composer a range of options when developing an abstract narrative between sound and image. The final type of montage, *Intellectual montage*, is intended to create overtones of an intellectual sort. The images may hint at something outside of the immediate imagery to an intellectual realisation or understanding of the film or its narrative.

In visual music, there can be a tendency for narratives to be assembled in accordance with intellectual type montage, but all other types of montage are powerful tools in composition and in creating relationships between the music and the image. As suggested above, rather than some extraneous concept informing the montage, in visual music one always has the music soundtrack to influence the decision-making process. Murch considers the emotional thrust of the storyline to be the priority when editing and I agree with this approach. It is not always an easy task when working with abstract materials but creating an emotional connection to a composition is one that will form a lasting impression in the viewer.

Composer Rendition and Algorithmic Techniques

The composition scenarios and techniques discussed above mostly involve all aspects of composition being undertaken by the composer in a process of *rendition*. Working with a concept, technologies, materials, a video or a soundtrack the composer arranges compositional elements until they form the desired outcomes. An alternative approach could be to use *algorithmic* techniques, whereby a computer or hardware system is programmed and utilised by the artist to create the desired results. By designing a process which can synthesise media, map between domains or create arrangements, a composition can be rendered entirely by machine and algorithm. Although presented as binary solutions to the same 'problem', these two alternative methods do not have to be employed exclusively. Algorithmic processes can be used to generate a range of media materials which can then be evaluated and incorporated by the composer via the rendition approach. Conversely, if algorithmic techniques are predominantly used the program itself will play a more influential role. Such algorithmic processes are commonly programmed or defined by the composer, ensuring its outputs meet the desired qualities and audio-visual relationships. An approach somewhere between the two extremes of completely algorithmic or total composer rendition can be fruitful. Algorithms, generative and mapping techniques can be used to produce a range of materials after which the composer can intervene and freely develop the interplay between pictures and sounds considering emerging visual and musical relationships. This approach has the potential to benefit from the advantages of both methods with greater audio-visual cohesion, structural intricacy and more sophisticated compositions driven by composer experience and intuition. In another sense the generation of some materials by the computer provides a type of collaborative working between composer and machine.

A Critique of Composer Rendition versus Algorithmic Techniques

In algorithmic generative techniques both sound and image can be created by a computer. The algorithm, under control of the composer, also permits the creation of both media elements simultaneously. Compare this with the previously discussed visual and music inception approaches where media are produced individually and independently. Similarly, it is possible to use generative techniques in one medium and to compose the other medium using more traditional methods.

Depending on the type and complexity of the algorithm, the results it produces may be more or less satisfactory when compared to a rendition approach, where the artist works directly with the media materials. One limitation of the *Hue Music* algorithm, described in the next chapter, is that it cannot create complex temporal arrangements. It is a real-time generative sonification method that produces a gradually evolving soundscape, which is somewhat repetitive. However, it was decided to leave the program in its current state as the soundscape nature of the results are somewhat effective and the level of programming challenge required to increase its complexity was distracting from the rendition approach described above and which, to me, feels more natural. The program is also available for the community to develop as they see fit with some suggestions given later.

In general terms, although algorithmic techniques can be useful in creating entire compositions, they are more effective in generating media items for later use in rendition. In my own experience good results can be achieved this way but one must expect a certain amount of coding to achieve something acceptable. Of course, the degree of time investment will depend on the programming skills of the composer. A happy balance however seems to exist somewhere between the algorithmic approach and intuitive interpretation and rendition. Algorithms tend to produce results that initially seem satisfactory. However, after a while they can appear predictable or repetitive and require intervention to form materials into a coherent work. Additional compositional structure and complexity can be implemented through manual editing and arrangement. This has been a recurring theme in my practice whereby algorithmic processes have generated useful ideas and source materials, but the finer structural details and narrative progression have required composer intervention. This issue was discussed (Garro 2005) with reference to a continuum whereby the processes of composer rendition and interpretation occupy one side of a continuum and automated parametric mapping occupies the other. Garro discussed a series of examples using a range of approaches between these extremes, with varying qualitative results. Battey and Fischman suggested a successful mix of correspondences, and higher-order intuitive alignment occur somewhere near the middle of the continuum (2016, 73).

The bonding between video and music was realised in *Hue Music* by material transference between visual qualities and sonic ones. This was simple to achieve in that, once the algorithm was programmed, the transference was automatic. Compositions that completely eliminate algorithmic methods and are undertaken by the composer are created by operations performed with hardware or in multimedia workstations. This requires the interpretive processes and material transference to be undertaken by the composer. Development of an inter-media synergy requires a type of intervention made simpler for some using ready-made tools such as media workstations as opposed to developing their own processes via programming techniques. Media production platforms aid composition by allowing video and audio materials to be combined readily and changes to be made easily. They allow rapid visualisation and preview of the results which is not always possible in a coding environment. New edits and musical arrangements can be evaluated rapidly, allowing a degree of compositional improvisation once familiarity with the workflow is gained.

It is conceivable, however, to create purely algorithmic works of complexity and beauty. This is the case in Bret Battey's *Autarkeia Aggregatum* (2005a). In this piece both the visuals and music were created via programming methods. Battey's programs visually generate, 'a constantly transforming, massed animation of nearly 12,000 individual points in a high-definition visual field' (Battey 2006, 1) and musically create 'naturalistic and expressive glissandi and continuous-pitch melodies' (Battey 2005b). The level of detail coded within the programs ensures the rendered piece is of a high quality in both technical and aesthetic terms. Despite this, Battey's program will always produce works with a similar granular style. To maintain diversity and originality each new piece requires a new algorithm, or at least a modification of the original. This would also be the case for *Hue Music*. This rule-based approach to composition can be fruitful but, as experienced, it becomes more a computer programming exercise than a musical composition one and limited the intricacy of the compositions. I am sure any live coders reading this would likely disagree, but the flexibility and speed of modern multimedia workstations allow a very intuitive approach to music and video composition and ideas can be developed quickly. It is up to the composer to decide the best approach and set of tools to use for any new composition and preferred working methods will be developed through experience. At the outset, each approach will begin with a blank canvas upon which the library of materials or lines of code are to be assembled and arranged.

Visual Music Composition Analysis

The preceding techniques represent a range of tools and processes which can be utilised to aid composition. It is also possible to retrospectively analyse existing works with respect to these methods providing insight and inspiration for new compositions. There is a limited toolset available to undertake analysis of visual music works and current discussions often take cues from music and film studies. Attempts to add to the discourse specific to audio-visual artwork have been undertaken by Harris (2021), who provides an in-depth analysis of several pieces from a variety of perspectives such as their shape, colour, time and space. Here I will focus on lumia factors, to analyse and illuminate two pieces but also take into account the previous discussions on audiovisual landscapes and gesture and texture. The intention is to encourage analysis and discussion within the visual music community.

Analysis of Aquatic by Hiromi Ishii

From a visual perspective *Aquatic* (Ishii 2016) is principally a study of evolving forms rendered with a minimal, subtly shaded monochromatic colour palette. It commences with a small group of particles in the corner of the screen which gradually rotate and expand. At first, they form cloud-like structures but soon start to resemble flowing waves and water. The colour palette consists of a green/white gradient throughout the piece, reflecting the underwater qualities, which are a central theme of the composition. This choice of colour palette successfully captures the essence of the subject matter, water and waves. The nebulous forms move and reposition across the screen with very subtle motion and transform between various states and shapes. The wave-like nature of the forms becomes more evident as the composition progresses, but the motion remains subtle throughout and reflects the qualities of slow-moving water. The soundtrack is very textural and drone-like. It appears synthetic but clues to its origin are revealed when water bubbles and ripples can be heard. The sound palette actually consists of recordings of whale song which are heavily processed. The sound forms are long and evolving, much like the visual forms. They are saturated sound textures with minimal gestural content and are coupled with similarly saturated visual colours. The soundscape is very spatial in nature and envelops the auditory senses as if to submerge the viewer under the water. Gradual sound motion is introduced by subtle sound panning. *Aquatic* straddles the line between a virtual landscape and a non-real one. The sound and images are created by heavy processing carried out on sound recordings and photographs which are placed in a virtual landscape but when viewing the piece, it gives the impression of a non-real landscape.

Analysis of Synthetic Electro Replicant by Dave Payling

Applying a similar analysis to my own work, *Synthetic Electro Replicant* (Payling 2016) reveals a different emphasis of factors. This composition is split into four distinct sections, which gradually increase the degree of synchronisation between sound and image. The opening 1'47" is a combination of textural drones and subtle rhythmic gestures. There is some synchronisation of audio and visual gestures, but the music is also complimentary in nature. During the introduction the visuals consist of abstract moving shapes above a textural coloured background. Moving into second section, sound and image become more rhythmically focused. Animated lines dance in syncopation with the percussive musical rhythm. The integration of rhythm into this piece created sequences that I conceptualised as 'audiovisual syncopation'. This was created by offsetting the synchronisation between sound and image. This 'offset-synchronisation' creates tension and anticipation between visual and auditory expectations. This is a visual motion coupled with a type of synthetic metric montage. To signal the transition into the third section at 3'03" a sonic gestural click signifies the switch of background colour from a greenish tint to a yellow one. This section properly begins at 3'22". The percussive rhythms are replaced with a keyboard line mapped to an abstract mandala form created from multiple overlapping geometric shapes which pulse to the keyboard melody. Material transference is in evidence here with visual motion mapped to the keyboard melody. This section closes out at 4'50" with a dimming of light and atmospheric background music. Moving into the final section a blue hue textures the background and the percussive rhythms completely dominate the soundtrack; another series of animated overlapping geometric forms are the central visual components. Colours are used expressively and are heavily saturated to accentuate the upbeat nature of this passage. Colour was used throughout this composition to attempt to create an emotional connection with the video and as a means of giving each section its own character. Different colours were chosen to emphasise the qualities of each section and are edited in an extended type of metric montage. Towards its conclusion the composition gradually breaks down into textural sound recapitulating the introductory themes. Overall, this piece is dominated by objects in motion synchronised to rhythmic gestural sound. Various shape forms including lines, hexagons and circles move and transform in a syncopated manner with the sound. *Synthetic Electro Replicant* is both visually and sonically a non-real landscape; there are no identifiable real visual spaces visible or sonic landscapes audible.

Notes

1 There is also a fourth technique related to the use of generative algorithms where both sound and image are created simultaneously.
2 This piece can also be considered in relation to metric and rhythmic montage.
3 Shepard is using chroma here to refer to musical pitch. It is not the same as Wilfred's chroma, which is a sub-factor of colour.

References

Adkins, Monty. 2021. *With Love. From an Invader*. Crónica. https://cronica.bandcamp.com/album/with-love-from-an-invader.

Adkins, Monty and Yan Wang Preston, dirs. 2021. *With Love: From an Invader*. Digital Video. Portugal: Crónica. https://pure.hud.ac.uk/en/publications/with-love-from-an-invader.

Barrass, Stephen. 1997. 'Auditory Information Design'. Australia: The Australian National University.

Battey, Bret. 2005a. *Autarkeia Aggregatum*. Audiovisual Composition. https://vimeo.com/14923910.

———. 2005b. 'Bret Battey-Gallery'. www.mti.dmu.ac.uk/~bbattey/Gallery/autark.html.

———. 2006. 'Autarkeia Aggregatum: Autonomous Points, Emergent Textures'. In *ACM SIGGRAPH 2006 Sketches*, 92.

Battey, Bret and Rajmil Fischman. 2016. 'Convergence of Time and Space: The Practice of Visual Music from an Electroacoustic Music Perspective'. *The Oxford Handbook of Sound and Image in Western Art*, August. https://doi.org/10.1093/oxfordhb/9780199841547.013.002.

Blackburn, Manuella. 2021. *Microplastics*. Stereo Fixed Medium. Electrocd. https://electrocd.com/en/oeuvre/49316/manuella-blackburn/microplastics.

Chion, Michel. 1994. *Audio-Vision: Sound on Screen*. Columbia University Press.

———. 2009. *Film: A Sound Art*. New York: Columbia University Press.

Cooper, Max and Kevin McGloughlin. 2020. *A Fleeting Life*. www.youtube.com/watch?v=ueFvV7QA9Kc.

Cooper, Max, Kevin McGloughlin, and Tom Hodge. 2017. *Symmetry*. www.youtube.com/watch?v=RG_TfS7ZopY.

Cope, Nick and Tim Howle. 2011. *Flags*. Electroacoustic Movie. https://vimeo.com/57190789.

Eggeling, Viking. 1924. *Symphonie Diagonale*. Animation.

Eisenstein, Sergei. 1949. *Film Form: Essays in Film Theory*. Translated by Jay Leyda. A Harvest Book 153. New York: Harcourt, Brace & World.

Field, Ambrose. 2000. 'Simulation and Reality: The New Sonic Objects'. In *Music, Electronic Media, and Culture*, edited by Simon Emmerson, 36–55. Aldershot, Hampshire: Ashgate Publishing Limited.

Fischinger, Oskar. 2004. 'A Document Related to R-2'. In *Optical Poetry: The Life and Work of Oskar Fischinger*, edited by William Moritz, 176–78. Bloomington: Indiana University Press.

Garro, Diego. 2005. 'A Glow on Pythagoras' Curtain. A Composer's Perspective on Electroacoustic Music with Video'. In *EMS International Conference Series, Sound in Multimedia Contexts*. Montreal.

Grant, Dwinell. 1943. *Color Sequence*. Animation, Short.

Harris, Louise. 2021. *Composing Audiovisually: Perspectives on Audiovisual Practices and Relationships*. 1st edition. New York: Focal Press.

Hyde, Joseph. 2009. *End Transmission*. Video Music. https://vimeo.com/3241660.

———. 2012. 'Musique Concrète Thinking in Visual Music Practice: Audiovisual Silence and Noise, Reduced Listening and Visual Suspension'. *Organised Sound* 17 (02): 170–78.

Ishii, Hiromi. 2016. *Aquatic*. Visual Music. https://vimeo.com/156292151.

Jarman, Ruth and Joe Gerhardt. 2010. *Heliocentric*. Video Music. https://vimeo.com/8129736.

Knight-Hill, Andrew. 2020. 'Audiovisual Spaces: Spatiality, Experience and Potentiality in Audiovisual Composition'. In *Sound and Image: Aesthetics and Practices*, 49–64. London: Focal Press.

Krause, Bernie. 2007. *Autumn Day in Yellowstone*. Field Recording. Wild Sanctuary. https://wildstore.wildsanctuary.com/collections/soundscape-albums/products/autumn-day-in-yellowstone.

Kuehn, Mikel. 2020. *Dancing in the Ether*. Electroacoustic Music. Soundcloud. https://soundcloud.com/mikelkuehn/dancing-in-the-ether.

McLaren, Norman. 1978. *Animated Motion: Part 5 by Norman McLaren, Grant Munro—NFB*. Animation. National Film Board of Canada. www.nfb.ca/film/animated_motion_part_5.

Miller, Dennis and Moon Young Ha. 2008. *Amorphisms*. Audiovisual Composition.

Moritz, William. 2004. *Optical Poetry: The Life and Work of Oskar Fischinger*. Bloomington: Indiana University Press.

Murch, Walter. 2001. *In the Blink of an Eye*. 2nd ed. Los Angeles: Silman-James Press.

Newman, Barnett. 1950. *Vir Heroicus Sublimis*. Oil on Canvas. Museum of Modern Art, New York.

Oliveira, João Pedro. 2018. *Neshamah*. Visual Music. https://vimeo.com/255169571.

Payling, Dave. 2016. *Synthetic Electro Replicant (Edit)*. Electronic Visual Music. https://vimeo.com/176765679.

———. 2017. 'Lumia and Visual Music: Using Thomas Wilfred's Lumia Factors to Inform Audiovisual Composition'. *EContact! Online Journal for Electroacoustic Practices* 19 (2). https://econtact.ca/19_2/payling_lumia.html.

Payling, David. 2014. 'Visual Music Composition with Electronic Sound and Video: Thesis Submitted in Partial Fulfilment of the Requirements of Staffordshire University for the Degree of Doctor of Philosophy'. Stoke-on-Trent: Staffordshire University.

Piché, Jean. 2017. *Threshing in the Palace of Light*. Video Music. https://vimeo.com/243462742.

Shepard, Roger. 2001. 'Pitch Perception and Measurement'. In *Music, Cognition and Computerized Sound*, Cdr Edition, edited by Perry R. Cook, 149–66. Cambridge, MA: MIT Press.

sisman, candas. 2010. *F L U X*. Audiovisual Composition. https://vimeo.com/15395471.

Sloane, Patricia. 1989. *The Visual Nature of Color*. New York: Design Press.

Smalley, Dennis. 1986. 'Spectro-Morphology and Structuring Processes'. In *The Language of Electroacoustic Music*, edited by Simon Emmerson, 61–93. London: Palgrave Macmillan.

Stein, Donna, M. 1971. 'Thomas Wilfred: Lumia'. *Press Preview*, 8 September 1971.

Wilfred, Thomas. 1947. 'Light and the Artist'. *The Journal of Aesthetics and Art Criticism* 5 (4): 247–55.

5 Create

Electronic Visual Music

Creating Electronic Visual Music

This chapter consolidates the ideas discussed previously and channels them into practical techniques for realising visual music productions.[1] It is not intended to be a prescriptive formula to be precisely followed, but more as a collection of suggestions which can be adapted, refined or discarded. Hopefully it will provide some useful pointers to get you started or further expand your skills. It introduces mostly computer-based techniques for creating visual music productions. My own experience is strongly influenced by using computers for making music and video, and all the benefits of accessibility to software tools that has afforded. The visual music computer journey began for me in the early 1990s after getting hold of an Amiga A500 which, as a consumer product, had very good graphics capabilities for the time. After starting out with Music-X, an early Amiga alternative to Cubase which was widely used on the Atari, it wasn't long before I was introduced to comparable visual software. Deluxe Paint (by Electronic Arts, the prolific games developer) was a graphics package which, among other things, allowed manipulation and animation using scanned photographs. This, in conjunction with Brilliance (another graphics package) and the Music-X MIDI sequencer, meant it was possible for an artist to create all the materials needed for live performance or recorded media. This was my first step in working with visuals alongside music and was something that really made the band, I was in at the time, stand out against others in the local area. Back then the sound, which was mostly electronic drums and programmed synthesiser tracks, was transferred to DAT tape and the visuals to VHS tape. No synchronisation was used during performance, the visuals were a free-running backdrop to the music performance, with live musicians and electronic backing. After several years using the Amiga I moved over to PC and more recently over to MacOS. With all the advances of computing and software technologies it is now possible (although not always recommended) to run real-time synchronised visual music performances using the same machine. Currently I use a combination of both Mac and PC, in what I feel is a best of both worlds approach. I like to use the PC workstation for heavy duty video rendering, as it is a more affordable route to using a powerful graphics card. The MacBook laptop is great for live performance duties and working with ideas while on the move. Most of the apps I use are cross-platform, meaning I can

DOI: 10.4324/b23058-6

share projects between machines. There can be compatibility issues with plugins and making sure each computer is using the same version of the software, so I do keep switching to a minimum. Some programmes are platform-specific, so it has been useful to have both Mac and PC available to give access to the most software. It is entirely possible however to create compositions using a single platform, and even a single application. The box in this section lists some software applications that are useful for visual music production.

Audio-Visual Software Tools

This is a list of apps I have used regularly for audio and video production. It is not an exhaustive list and more suggestions can be found in the interviews in Chapter 2.

Touchdesigner

Alternatives. Vuo, Resolume Wire, Notch, Max (Jitter)

https://derivative.ca Try the free non-commercial license before deciding if you need the additional features of the paid versions.

Touchdesiger (TD) is a cross-platform node-based programming tool with some similarities to Quartz Composer. The non-commercial version is free to use but has some limitations, such as reduced resolution, when compared to the paid versions. TD is great for creating real-time interactive patches and allows video to be combined with digitally generated images quite freely. There is a wealth of video tutorials online, those from bileam tschepe (elekktronaut) (www.youtube.com/c/bileamtschepe) being an excellent introduction.

Quartz Composer

http://developer.apple.com/downloads Search for Quartz Composer or Additional/Graphics Tools for XCode. Will work best on older versions of MacOS up to 10.10 Yosemite.

Quartz Composer gets an honorary mention as it is now an unsupported application but has been a very useful tool. It is a Mac-only app, which uses OpenGL and since Apple stopped supporting OpenGL and switched to their alternative, Metal, Quartz is no longer supported. It is, however, a free programme still available from the Apple developer website and can run on older versions of MacOS. It is a node-based programming tool which can create interactive patches generating high-quality motion graphics, image effects and abstract animations. Quartz patches can also be converted to plugins using the paid application FX Factory, for use in Adobe After Effects and other graphics apps. Most of my animated compositions between 2014 and 2018 were created this way.

Adobe After Effects

www.adobe.com/uk/products/aftereffects.html

After Effects (AE) is a very powerful tool for visual effects and motion graphics. I think of it as a cross between Adobe Photoshop and Adobe Premiere, as it works with video layers, blend modes and effects similar to Photoshop, while having more comprehensive video functionality. Video editing in AE is somewhat cumbersome, so if you intend to work on a video-only composition consider using Premiere or DaVinci Resolve. AE is very good for keyframing and animating transitions and effects parameters. It also possible to import audio stems and generate keyframes from their amplitude values which can then be mapped to visual parameters. This technique of parametric mapping can create subtle and not so subtle results. AE can also be used for post-processing of otherwise completed videos with colour and other effects and for generating animated title sequences.

DaVinci Resolve

www.blackmagicdesign.com/products/davinciresolve

As well as being a powerful video editing and compositing tool, Resolve is actually a fully featured audio-visual application allowing colour grading, effects keyframing and audio post-production. A cross-platform app, there is a free version with a few limitations, while the paid version has more effects and higher resolution capabilities. For video-only productions, those not using abstract animations, this would be my recommended tool.

Resolume

Alternatives: VDMX, Modul8

https://resolume.com/

Resolume is a real-time graphics application for performing live visuals. It comes in two flavours, Avenue VJ Software for AV performance and video effects and Arena, which has additional capabilities for large venues such as advanced projection mapping and projector blending. Resolume can be used to apply effects to, and control source playback of, visuals generated by other applications such as Touchdesigner using either Syhpon on MacOS or Spout on PC.

Notch

www.notch.one/

Notch is a relative newcomer to visual production but is possibly the most powerful, fully featured and optimised app for video effects, motion graphics and real-time visuals. Like many of the apps above, Notch is a node-based programme but it also has timeline capabilities for keyframing and composition all within the same interface. Its particle features are particularly impressive. Notch also has VR capabilities. The main downside to Notch is its cost, but it has a less expensive 'learning' license.

Max

Alternatives: Audiomulch, Pure Data (PD)
https://cycling74.com/
Possibly the most comprehensive and mature node-based programming platform for audio and MIDI, Max is cross-platform and can be used on both Mac and PC. Max has an extensive user base and an integrated system of tutorials. It is great for creating stand-alone interactive apps for MIDI and audio processing and synthesis and runs in real-time. It also has graphic manipulation and generation tools making it possible to create integrated audio-visual applications with a single interface. While being excellent for control and processing, its sequencing capabilities are somewhat weaker. This is overcome however by being able to run Max patches inside Ableton Live, using the Max for Live package. This makes Max an extremely powerful application. Max's open-source relative, PD https://puredata.info/, has similar functionality and a dedicated user base.

Ableton Live

www.ableton.com/en/
Ableton can be used both as a performance tool and a sequencing and composition workstation. Its real-time options allow clips to be launched and synchronised similarly to how video clips are launched and managed in Resolume. It can be connected to Touchdesigner and other apps making it possible to run a video app on one machine and Ableton on another while keeping everything tightly synchronised. Ableton also has an extensive suite of plugins for MIDI and audio processing as well as a comprehensive range of synths and samplers with the greatest number of options contained in the Ableton Suite version.

Steinberg Nuendo

Alternatives: Logic Pro, Pro-Tools, Cubase, Presonus Studio 1
www.steinberg.net/nuendo/
Nuendo is my (not so) secret weapon when it comes to music composition and production. It is the more fully featured sibling of Cubase with additional post-production features including Dolby Atmos support. Initially developed as a digital tape machine, akin to Avid Pro-Tools, it has grown to be one of the most comprehensive digital audio workstations around.

Native Instruments Reaktor

www.native-instruments.com/en/products/komplete/synths/reaktor-6/

Reaktor is a very powerful synthesis and effects tool. Essentially it is a node-based audio and MIDI programming toolkit, similar to MAX, but it has a library of prebuilt synthesisers and effects that is ever expanding and is generally a more effective app and plugin for high-quality software synthesis. More recently it has included a series of modular devices which can be used and programmed as a virtual modular synthesis rack.

Visual Music Production Workflow

The term 'production' in this context refers to the overall process of preparation, creation and finalising of a visual music composition which primarily involves the stages of:

1 Recording and preparation
2 Arrangement and composition
3 Mixing, mastering and rendering for delivery format

These stages apply to music and video composition individually, and to the combined media. One might progress through these phases completely to produce the music before commencing work on the visuals, for example. The phases are not necessarily linear and are often recursive, requiring several iterations to finish a piece. A composition may seem complete but often on post-render viewing and evaluation things can appear out of place, missing or just wrong. Posting a video to an online sharing site is a very effective way of revealing issues that were not previously noticed. If changes are needed at this stage, it may be necessary to return to recording, prepare new material and further develop the composition. There is also interaction between all three stages, especially in electronic music composition. If new sound elements are created in stage 1 and inserted into the arrangement at stage 2, this can force further compositional work, requiring a new render. Seemingly trivial decisions can therefore result in a structural re-edit or complete overhaul of the composition.

Csikszentmihalyi (2013) suggested there are five steps to creation, namely *preparation, incubation, insight, evaluation* and *elaboration* and this is a good framework which can be applied to visual music production. When considering how these five steps relate to the three stages above, *preparation* takes place in stage 1, *incubation* and *insight* in stages 1 and 2, *evaluation* in stage 3 and *elaboration* again in stages 1 and 2. It is often necessary to continually elaborate the developing production, driven by self-evaluation and the composer's intuition. Do not underestimate the need for an incubation period. Leaving a project partway through allows the ideas to mature subconsciously and gives oneself a renewed objectivity when returning. One advantage of working with multiple media is that time spent on the visuals provides a break from the music and vice versa, allowing time for incubation and hopefully an 'aha' moment when the idea becomes more crystalised.

Stage 1   Recording and Preparation

Preparation for music composition can begin with the collection of sounds and creation of a sound library. This is typically the process when working in electroacoustic genres, as Tim Howle discusses in the interview in Chapter 2. A sound library, comprised from field recordings, is a valuable resource to develop and can be gradually accumulated over time. Unique sounds are best recorded by the artists themselves, rather than using sample libraries, and can be captured using a portable field-recorder. There are many inexpensive devices now available for this purpose and other more expensive options that can be coupled with high-quality microphones. Field recordings are necessary for capturing sounds which occur naturally in the environment and cannot be replicated in the studio. Do consider using a studio to capture detailed sounds, but field-recorders allow an immediacy to capture sounds that may suddenly occur, like a person running past, opening a gate and running off into the distance. Being able to capture sounds in an opportunistic way like this creates individualistic sounds and those that could not be recreated in other ways. There is also an inherent spatial quality to field recordings (with devices with more than one channel) which is created by the environment itself: the ambience of the room or the outdoor natural soundscape with subtle environmental sounds in all directions can add an extra dimension to a piece of music. Soundfield microphones, although expensive, are particularly adept at capturing environmental space (Soundfield 2023). It is also possible to move the position of a portable recorder itself in real-time to track movement or change the spatial positioning of sounds, but beware of handling noise, best identified by using headphones during recording. If an interesting sound is discovered, such as the resonance created when tapping a particular section of pipework, it is good practise to record several versions of the same sound using different microphone positions and to minimise (or have varied) background noise disturbances. The best takes can then be chosen and saved in editing. Of course, as suggested, for precise control some sounds are best recorded in a recording studio or other acoustically optimised environment to capture the most detail and clarity possible.

Once the source materials are captured, development of the library can commence. For this task the sound files will require auditioning and trimming using a dedicated audio editor such as Adobe Audition, or an equivalent tool like Audacity. These offer specialised tools for this task and will also allow sample rate conversion, noise reduction and other 'destructive' type processing but a similar task can be done in a DAW. Do not discard sounds which initially seem inappropriate for the production currently in progress. They will potentially become useable, even years later. This preparatory work of editing, treating and categorising sounds can be time-consuming and potentially mundane, but will speed the workflow when arriving at the arrangement phase. I forget the source, but this quote seems particularly relevant in this context: 'it is in the preparation for the work that the work is actually done'.

When creating the library consider different categories of 'instrument' that may typically be found in musical compositions. They might comprise of short percussive elements, longer evolving textural sounds, and so on, and others designed to inhabit specific frequency ranges such as bass parts. The original recordings can also be enhanced with various spectromorphological processes at this stage to enhance their musical qualities and identification with the desired sonic and audio-visual landscapes as discussed in Chapter 4 and those being developed in your composition. Even if you are working in other electronic genres, including beat-based EDM styles, found sound layers can add another level of intrigue to the mix and are worth considering.

Synthesised sounds are produced in an entirely different manner and a full discussion here is not possible. A library of synth sounds can be pre-produced but a major difference, and advantage with synthesis, is that the desired sound qualities can be specified through programming. Sound synthesis presents the opportunity to adjust the overall timbral and dynamic character of a sound by adjusting filters, modulation and other parameters so they fit with the developing composition. Synthesisers can also be performed in real-time alongside the visuals to exactly produce the desired results and to gel cohesively with the visual narrative. Similar to above, even if the composition is comprised mostly from found sounds, some synthesised sounds can be useful for creating specific sonic qualities not available in the found sound library.

Film Footage Recording

Capturing film footage can be done in a similar way to audio field recordings. Depending on the desired quality, video can be recorded on mobile phones, high end cinema cameras and everything in between. Similar to audio, having a portable video recording solution allows the capture of opportunistic video footage, and there are many devices suitable for this, including the Go-Pro action cameras and touch-screen phones and tablets. Sometimes, however, a high-quality camera mounted on a tripod is the best solution for capturing high-quality footage of a specific scene. When filming and building a video library also consider how multiple videos layers will combine. Static visual textures can combine well with gestural, movement-based, videos. Also consider the colour content of a scene and how this will contribute to the resultant quality of the compiled visuals. Like sound it is also possible to synthesise visuals, with some techniques being discussed later.

Stage 2   Arrangement and Composition

The arrangement of a composition can begin when the sound library contains enough contrasting and related sounds to allow the music to begin taking shape or when the composer is ready to begin working on the synthesised and sampled parts. One solution to this consists of arranging sounds and parts horizontally in time and vertically as tracks inside a digital audio workstation (DAW). Layering multiple sound layers 'vertically' permits a complex spectral picture to be created and arranging them horizontally across a timeline produces the structural form of a composition. This method of working can be thought of as a bottom-up approach with: 'works based on materials they (the composers) have assembled which they subsequently manipulate and place in sequences to form structures' (Landy 2007, 34). This requires an intuitive approach; gradually expanding and refining the composition as it develops. The alternative to bottom-up composition is a top-down approach. This involves deciding a predefined structure, or compositional form, that influences the arrangement process. Top-down structures are sometimes imposed and transferred between media; when either the video or music are completed in entirety and the structure of the other medium must follow for example. This is a visual or musical inception process as described in Chapter 4. Even if it was not determined initially, it will be necessary at some point to consider the overall structure and flow and form of the composition. In reality a combination of bottom-up and top-down methods is used as a composition develops. Of course, if you are composing a video you have an additional guide, a top level structure and set of constraints with which to work creatively.

Similar to the compositional processed used to create the music, the moving image in a project can be compiled and arranged from a video library. Video clips can be placed horizontally and vertically on a timeline to create the desired flow through the composition. There is such a strong crossover workflow between DAWs and non-linear video editors (NLVEs) that the video composition process is effectively the same, except the composer is creating a visual rather than an auditory flow.

Conversely to the above compositional process taking place in a DAW, an electronic artist using synthesis hardware or plugins could perform much of the composition in real-time. With a collection of connected synths, drum machines, effects and modular devices it is possible to create all the music in one take. The laptop artist could take a similar approach using Ableton Live or an alternative platform. The music performance can be done after the visuals are finished whereby the artist can respond to the visual narrative by improvising to the video recording. Good practice would be to record each instrument into its own track in a DAW. This permits further elaboration, processing of individual instruments and editing the structure of the composition, which can be trimmed and rearranged. Multiple takes of the same performance can also be comped together to capitalise on the best bits.

Stage 3   *Mixing, Mastering and Rendering for Delivery Format*

These are the final stages in the production process. The flexibility of DAWs and NLVEs means compositional arrangements can be modified all the way until the final master is rendered. Sound mixing is a skill which can be developed and whose processes are outside the scope of this book, but there are many good reference sources available. Mastering is another art, this time at the final stage in audio production, and involves using any equalisation, dynamic and enhancement type processes as required to produce a clear, detailed and dynamic piece of music as possible before being rendered for later use with the video. It is good practice to create the music track in as high a resolution as possible. A minimum high-quality standard as of writing is two channel, 96KHz sample rate and 24bit (word length) resolution. This high-resolution master can be down-sampled to lower resolutions where necessary, but no quality is gained from up-sampling a lower resolution file to a higher one.

In terms of dynamic range it is recommended to use a loudness standard derived from ITU-R BS.1770 (ITU 2012). Many video and music streaming platforms will use a variant of this standard to normalise audio levels and ensure consistent loudness between program materials. When played at a concert the sound engineer will manually adjust levels to maintain consistent loudness between compositions. Therefore, there is not always something to be gained by making the soundtrack as loud as possible, as it will potentially be turned down during playback. There are several loudness meters that will ensure compliance with ITU standards, such as the YouLean Loudness Meter (YouLean 2023). A common loudness level used by Spotify and YouTube is -14 LUFS. This is a good loudness level and will preserve the dynamics of the music while still being relatively loud. For even more dynamics consider using the EBU standard of -23 LUFS. This will permit a large variation in loudness throughout a composition allowing subtle sections to be contrasted with more impactful passages.

For multichannel works Dolby Atmos is one technique to allow mixes to be scaled to any available speaker configuration. If working to a known loudspeaker layout, often stereo for videos, it could be more appropriate to configure the DAW project in that specific channel-based format to ensure maximum compatibility.

For the finished video, render in a suitable file format such as QuickTime 1920 x 1080, H.264 compression codec, 25 fps video, 16bit 48KHz PCM audio (or higher). This should be considered a minimum standard. Data rates between 20,000–30,000 kbps will produce a good trade-off between quality and file size. Higher data rates will preserve more detail and are good for creating a high-resolution archive master file. Frame rates up to 120 fps will produce smoother animations and can be useful when feedback-type effects are used, as the feedback decay will appear smoother. Depending on platform and detail in the video, higher data rates and/or higher frame rates can cause very long render times. This can be in the order of several hours, rather than the few minutes if might take to bounce an audio file from a DAW.

Image and Sound Terminologies

There are a variety of terms related to colour, sound and music that are used when composing, performing and using audio-visual tools. Those useful to this discussion are described here.

Colour

Colour can be classified in several different ways. Additive colour mixing uses the primary colours of red, green and blue (RGB). This type of mixing uses RGB coloured lights which can be added together to create other colours. This type of colour mixing is used in computer monitors and other displays. This is a common classification when working in audio-visual platforms. Subtractive colour mixing on the other hand is used in printing, and it uses cyan, magenta, yellow and black (CMYK) to create other colours. CMYK colour mixing uses pigments which effectively subtract, by a filtering process, other colours from light. A pure yellow pigment leaves only yellow visible and filters out the others. In both RGB and CMYK colour mixing, any other colour can be created from the primary colours. In RGB the production of white requires all three colours at equal strength but in CMYK printing white is created from an absence of pigment on a white background.

An alternative to these two systems is the HSL colour model. This specifies colours in terms of their *hue, saturation and lightness* (HSL). There are similar systems that replace lightness with either 'value' or 'brightness'. These are known as the HSV and HSB colour models respectively. The three HSB parameters have been described as:

> *Hue* indicates a particular colour sensation which is dependent simply on the relevant wavelength; the inherent colour of a thing; the purest or brightest form of a colour having no white or black mixed with it. A particular colour or colour name. Each hue has an intrinsic tonal value on the chromatic scale.
>
> (Paterson 2004, 203)

> *Saturation*. The intensity or purity of a hue; the extent of its colourfulness; the strength or richness of a colour indicating whether it is vivid or dull. The colours of the greatest purity are those in the spectrum. A colour with a very low purity is on the verge of becoming grey.
>
> (Paterson 2004, 349)

> *Brightness*. The condition of being bright. The value or luminosity of a colour. Yellow has the highest value in the spectrum and violet is the darkest in the spectrum. The two extremes of brightness are, of course, black and white. The addition of black or white to a hue changes its brightness.
>
> (Paterson 2004, 68)

Compare these definitions with Wilfred's Lumia colour factors, which were discussed in Chapter 1. For artists using a palette of paints, a starting pigment is selected which can then be lightened by the addition of white or darkened by the addition of black. In watercolours, saturation can be reduced by the addition of more water to make the hue less vibrant. For an artist therefore, HSB/HSL is a more intuitive way of conceptualising colour. In computer applications it is common to be presented with a digital 'rainbow' palette which allows rapid selection of the required colour. The chosen colour can then be finely adjusted by changing the RGB and HSL parameters, often accessible in the same window.

Sound

Just as colour can be classified by several parameters, there are four properties of sound that are commonly used in its description. The first of these is loudness. Humans can perceive very large differences in loudness from an extremely quiet whisper to a jet engine roar. Loudness is best measured in a 'loudness units full range' (LUFS), as this is a commonly used standard that takes into account the psychoacoustic properties of the human hearing system. Another property of sound, as produced by conventional musical instruments, is its frequency, note or pitch. In stringed instruments this relates to the rate at which the strings vibrate. Faster vibrations create a higher pitch and slower is lower. A third property is a sound's duration or note length. A sound cannot exist without occupying a finite amount of time and 'if the amplitude of the sound changes over its duration, the curve that the amplitude follows is called the amplitude envelope' (Roads 1996, 95). Percussion instruments often have a very loud and fast initial transient followed by a slower decay back to silence. The amplitude of a sound will vary over its duration and can be described by its envelope and each sound and instrument will have its own characteristic envelope.

The last, and perhaps least easy to quantify quality of a sound, is its timbre. Timbre is from the French word meaning stamp, in this case referring to the 'acoustic stamp' or tonal quality of a sound. Another way timbre can be described, is that it is the property that distinguishes one instrument from another when playing a note at the same pitch. Middle C on a piano will sound different to middle C on a church organ for example.

Acousmatic Sound and Electroacoustic Music

The parameters above describe sound in basic terms. In electroacoustic music further concepts are used, such as reduced listening, which form part of a wider language used in its discussion:

> The acousmatic situation changes the way we hear. By isolating the sound from the 'audiovisual complex' to which it initially belonged, it

creates favourable conditions for reduced listening which concentrates on the sound for its own sake, as sound object, independently of its causes or its meaning.

(Chion 1983, 11)

In reduced listening the cause of the sound is not necessarily considered when appreciating the music. The original sounds can even be transformed in ways to create surrogacy that further remove them from their causal origins. This is undertaken to remove existing preconceptions and create interest in the qualities of the sound itself. Spectro-morphology (Smalley 1986, 61) is a means of describing sounds processed and used in this way in electroacoustic compositions:

> Spectro-morphology is an approach to sound materials and musical structures which concentrates on the spectrum of available pitches and their shaping in time. In embracing the total framework of pitch and time it implies that the vernacular language is confined to a small area of the musical universe . . . it is sound recording, electronic technology, and most recently the computer, which have opened up a musical exploration not previously possible.

(Smalley 1986, 61)

Spectro-morphology therefore deals with the transformation of the pitch and timbral characteristics of a sound over time. The result of various transformations can produce changes in pitch, amplitude and other qualities producing complex sound objects of compositional interest. Although some of the traditional concepts of sound and music composition, such as the structural form of a piece, are still relevant, electroacoustic composition creates music of very different qualities to those created by traditional musical instruments. Although it is a genre of music to be appreciated for its own qualities electroacoustic music is paired to video by many audiovisual artists as it is well suited to the experimental nature of their videos. It is worth exploring at least some of these techniques for creating unique sound and image relationships but, of course, any type of music can be paired with video.

Video-based Visual Development

This section will explore video-based techniques using still images and film footage for creating materials to be used in composition. There can be several stages involved in creating suitable imagery for videos for a composition, but sometimes carefully shot original footage will be the most appropriate. The processes described below offer some techniques for enhancing and creating abstraction from original footage which can enhance the visual component of a composition.

Although visual music is often aligned with abstract animation, strong audio-visual cohesion can be achieved with real-world imagery.

Image Zooming and Scanning

This effect was used extensively by the documentary filmmaker Ken Burns as a means of imparting a temporal quality to still images. An example can be seen in the title sequence for *Civil War* (Burns 2012), in which photographs of soldiers and war scenes are used to create an evocative effect. Originally created with a real camera, it is possible to recreate this effect digitally. Find a suitable photograph, load it into the NLVE timeline, zoom in (enlarge) and automate a pan and zoom across the image. This can be done randomly, and some preset effects will do this automatically, but better results can be achieved by choosing visual elements of interest and deciding how to occasionally focus on these and move around the image to create the desired temporal and emotional results. Panning and zooming into digital photographs in this way can create low-resolution pixelated results and reduced video quality. These low-resolution artefacts can be reduced by applying smoothing effects such a blurring and other processing and also by using photographs with a high resolution.

Kinestasis

Kinestasis is another technique using still images which can create apparent motion. Instead of panning around a single photograph, several images are used in sequence to create a dynamic effect. Kinestasis has some similarities to time-lapse and stop motion photography but usually involves images of real scenery or figures. If several photographs are sequenced one after the other the result produced is similar to a low frame rate video and flickering of the images. An example of this type of kinestasis can be seen in *Circular* (Cooper 2020; McGloughlin and Cooper 2019). Compare this with the purer colour-based flicker results that can be seen in the films of Sharits and others (1966; 1968). Images to be used in kinestasis can be captured using video recording, with a low frame rate or by later omitting frames or individual photographs of a scene changing over time. When working with the images in the NLVE experiment with different frame rates for various results.

Video Transitions

Creating blended transitions between two photographs (or videos) is another method of treating and enhancing video footage. The effects created by transitions with different blend modes can produce visually appealing results. The way the shapes and their colours interact during different types of transition produces a wide variety of coloured hues and visual textures. Note that is it the images created during the transition, rather than a transition from one image to another, which is of most interest. Using only still images, the Ken Burns effect or kinestasis and transitions, visually intricate abstract visuals can be produced.

Symmetry in Video Production

Techniques involving symmetry can be another fruitful source of enhanced film footage. Source footage can be captured for this very purpose and with the intention of later applying symmetry. One approach is to scan the camera in a flowing movement across the source material while it is being recorded. This can create abstract unfolding patterns when the mirroring and related processes described below are applied and can be particularly apt for visual music composition.

Symmetry is useful as it imparts a degree of visual harmony where it may otherwise be absent. Symmetrical images can appear well balanced and proportioned and even reflect some of the harmonious qualities found in music, even when the original source material is somewhat mundane . . .

> the word symmetry is used in our everyday language in two meanings. In the one sense symmetric means something like well proportioned, well balanced, and symmetry denotes that sort of concordance of several parts by which they integrate into a whole. Beauty is bound up with symmetry. Thus Polykleitos, who wrote a book on proportion and whom the ancients praised for the harmonious perfection of his sculptures, uses the word, and Dürer follows him in setting down a canon of proportions for the human figure. In this sense the idea is by no means restricted to spatial objects; the synonym "harmony" points more toward its acoustical and musical than its geometric applications.
>
> (Weyl 1952, 3)

One type of symmetry is bilateral symmetry. This is where an image is symmetrical with respect to a plane (Weyl 1952, 4). This type of symmetry can be seen in nature, the left and right wings of butterflies, although not exact replicas are largely symmetrical about the centre line of the butterfly's body. In video and still images, bilateral symmetry can be created simply by reflecting, or mirroring, one half of an image onto the other half as demonstrated in Figure 5.1. When applied to moving images this creates an evolving pattern of lines and colours that appear to move outwards or inwards relative to the plane of reflection. If overused, the effect created by bilateral (and other types of) symmetry can become predictable and repetitive. Some variety can be achieved by offsetting the centre of the reflection plane or moving it dynamically.

Another type of symmetry is rotational symmetry. Here, a portion of an image is selected and rotated several times around a point. In Figure 5.1, the bottom right quarter of the bottom left image is repeated three times by rotating it 90 degrees about the centre of the image. It is also possible to select a smaller portion of the image and rotate it a greater number of times over fewer degrees for more complex

Figure 5.1 Top Row: Reflecting One Half of an Image in a Plane (E) to Create Bilateral
          Symmetry. Bottom Left: Rotational Symmetry, Bottom Quadrant Repeated and
          Rotated. Bottom Right: A Kaleidoscope Effect

symmetry. With additional mirroring this can create kaleidoscope-type effects.
Using symmetry in these ways distances the images from their original source in a
similar way that sound transformation can remove a sound from its original source
cause. It is a type of visual transformation. For abstract visual music this could
be considered beneficial as the movements and transformations of the moving
shapes become the focus of the videos rather than any representational imagery
they contain.

Synchronisation

A key factor of visual music pieces is the way the video and audio interact in a
temporal and synchronous way. In audio-visual terms, one result of synchronisation
is 'synchresis', defined as:

> the spontaneous and irresistible weld produced between a particular auditory
> phenomenon and visual phenomenon when they occur at the same time. This
> join results independently of any rational logic.
>
> (Chion 1994, 63)

Although these synchretic moments can be very short in duration, the effect they produce can persist throughout a piece, ensuring a continued bond between sound and image. Synchresis as a concept has been further explored and expanded (Boucher and Piché 2020) and Garro (2012, 107) also suggested that synchresis is only one point on a continuum, which extends on an axis from separation at one end to intuitive complementarity and synchresis, and culminating with parametric mapping at the other end of the axis. In this classification separation is the loosest form of synchronisation, where the soundtrack and video can be seemingly unrelated with no discernible moments of synchronisation. Observe *Open Circuits* (Cope and Howle 2003) from 5'40" for an example of this. Complementarity, the next step on the continuum, does not involve tight synchronisation but it is evident there is a temporal and qualitative relationship between sound and image. This can be observed in gradually evolving and flowing pieces such as Piché's *Océanes* (2011). Synchresis is achieved with a very close synchronisation and this can be achieved by parametric mapping, which is actually a technical process of mapping sound to image parameters or vice versa as described in Chapter 4. When using this technique, a gesture in one medium will create a corresponding gesture in the other, inherently causing a very tight synchronisation.

Although some pieces use one specific type of synchronisation, others will use multiple levels to create continued variation through the work. *Open Circuits* (Cope and Howle 2003) is a good example of differing degrees of synchronisation. The video was produced from several silent video clips that were mixed in real-time using analogue equipment. The video artist, Nick Cope, rapidly switched between the clips, creating a disorientating effect. The resultant video was recorded and passed onto Tim Howle who composed the music in response to the imagery. During the initial sections of the piece, the music responds to the quickly changing video with rapid edits between one sound and another. This naturally creates a tight synchronisation producing several synchretic moments. In the latter parts of the piece, the synchronisation breaks down and the music and video separate, moving towards complementarity, and ultimately separation. Another example, *End Transmission* (Hyde 2009), incorporates various levels of synchronisation which are mostly focused towards rigid synchronisation. This is due in part to the nature of its hardware hacking production techniques. Analogue equipment was used to produce, simultaneously, the sound and images and the results straddle the area between synchresis and parametric mapping. There are other passages with materials which are more complimentary in nature, creating contrasting qualities throughout the piece.

The use of various levels of synchronisation can therefore be an effective tool for creating temporal variety, tension and resolution in compositions. Consider creating passages where sound and image are more complementary in nature and contrasting these with intense passages of tight synchronisation to create sustained interest throughout a piece.

Fusion of Sound

Although synchresis, as described by Chion, relates to the fusion of sound and picture, a similar phenomenon can occur purely with sound. If two sounds occur at the same time they can become unified as a single sound 'object'. One way of achieving this is by treating the individual sounds with similar effects, so the qualities imparted by the effect dominate the original sound. For example, if two independent sounds are treated with the same intense form of amplitude modulation. Regardless of the tonal quality of the original sounds, because of the processing undertaken on them they become audibly fused as a single sound source. A similar phenomenon was described whereby . . .

> all three sounds now appear very closely related whereas the sounds from which they originated have no relationship whatsoever. At this distance of derivation it is the overall morphology of the sound structures which predominates.
>
> (Wishart 1996, 66)

Alternatively, multiple layers can contribute to a single sound. This is a common technique in sound design where 'body', 'click', 'scratch', 'thud' and other qualitative sound layers are created then mixed together to create a complete sound character for a chosen purpose.

Mixing Video Layers

DAWs and NLVEs share many capabilities in their workflows. Although this allows the musician or video artist to readily switch between production of the different media there is one key difference to how video and audio behave when mixing multiple layers together. If two musical instruments are mixed in a DAW by adjusting their relative amplitudes, depending on source material, it will appear that they are part of the same recording or performance. If, however, multiple video layers are made visible by adjusting their transparencies each video scene will maintain its own form and content and be perceived as separate scenes. In an image there can be certain familiar combinations of objects, such as a horizon line dividing the sky and terrain across the horizontal. If another video with a similar scene is blended with the first, both will maintain their perceptual independence with two obvious horizon lines. Similarly, scenes with unrelated content can blend together in an unconvincing manner, failing to produce the desired effect of creating a new, single, unified scene from the two separate layers, as would be the case with audio tracks. This effect is demonstrated in the middle image of Figure 5.2, where the two images are blended through transparency adjustments. This could in some part be resolved by making sure any original footage is captured in a way that the blend is successful.

Chroma-Keying

One remedy to this issue is chroma-keying, a type of compositing, whereby some elements of one image are replaced with another. This is commonly used in film

Figure 5.2 Blending versus Keying. Top Row: Original Images. Middle Image: Created by Adjusting the Transparency of the Landscape. Bottom Image: Created by Keying the Sky from the Landscape, Allowing the Statues to be Visible Above the Horizon Line

and television where people are recorded against a key colour screen so they can be later composited using chroma-keying against a new background. Chroma-keying is:

> A function that will render a specific color in a layer transparent. For example if you shoot someone in front of an evenly lit blue screen, you can use a chroma key function to render the blue completely transparent, thus revealing underlying video layers.
>
> (Long and Schenk 2011, 448)

A classic example of this is the presentation of a weather forecaster in front of a green screen so a weather map can be superimposed into the background. This is a conventional use of keying and, although it is common to use green or blue screens in film work, digital processing allows any colour to be chosen for this purpose. When this technique is used in a less conventional manner, an arbitrary object or background colour from one video can be chosen as the key colour. The non-keyed objects from that video can then be overlaid on top of other layers. If this process is applied to material that was not originally recorded with the intention of chroma-keying, for example in a complete moving scene which contains several objects, it can create interesting cut-out effects and the keyed image can appear to be part of the background image in the second video. This is a completely different result than that achieved by blending the videos using standard transparency techniques, but it is more analogous to results obtained by mixing sounds. The movement in one video does not need to match that of the other, the visual integration can still appear to be much tighter. The bottom image in Figure 5.2 shows the results of keying two scenes together using this method; it is possible to juxtapose entirely unrelated visual material but still make them combine more convincingly. Chroma-keying is therefore a very useful tool in creating video compositions for visual music.

Blend Modes

Results similar to, and potentially even more useful than, those produced with chroma-keying can be created by utilising different blend modes. The second image in Figure 5.2 demonstrates the effect created when a traditional crossfade is employed to reduce the opacity of each layer with a 'normal' blend mode. 'Normal' blend mode in video or image editing software obscures layers below the top layer unless the opacity of the top layer is reduced. By selecting a different blend mode, the pixels in each layer are combined using various arithmetic operators. For example, an additive blend mode will add each pixel together using the pixel values: a white pixel = 1.0 and a black pixel = 0.0 added together will produce a white pixel. Grey pixels have fractional values between 0.0 and 1.0. Figure 5.3 illustrates a selection of different blend modes, with two image layers at 100% opacity. In this case the top layer does not obscure that underneath, even when it is completely opaque. When applied to full colour images the results, particularly when working with abstract materials, can be very effective.

Figure 5.3 Example Blend Modes Combining the Landscapes in Figure 5.2. Top Left: Add.
Top Right: Colour Burn. Bottom Left: Difference. Bottom Right: Divide

Abstract Animation

Animation is an alternative approach to using filmed footage for creating the
visual component of visual music. I generally find that film footage is a more
streamlined way into composition, partly due to the accessibility of the necessary tools and the way production workflows translate between sound and image.
When using film stock, control is possible at the capture stage and selection in
editing and post-processing, but there are limitations to what can be achieved.
Animation, on the other hand, is an excellent technique for generating visuals that
contain all the necessary gestures, colours and other properties the artist desires.
It is akin to sound synthesis techniques in that it permits very precise control over
multiple parameters, the ability to structure them over time and also to interpret
relevant qualities in music, thereby creating enhanced audio-visual cohesion. The
downside with animation is that it will generally involve a long learning process
to acquire the necessary skills to produce results with the desirable qualities. In
some ways this issue is alleviated with the use of abstract animation. With abstract
materials it is not necessary to recreate complex natural behaviours or anthropo-
morphic animations that might be required to create life-like results. This frees the
artist to explore how musical qualities can be interpreted in abstract visuals and to
maximise the synergy between sound and image. Abstract animation is therefore
a popular type of visual (Kanellos 2018) paired with music in visual music and
audiovisual compositions.

Approaches to abstract animation include the use of particle systems and primitive shapes such as lines and squares. Although initially simplistic in nature, simple shapes can be warped, twisted, replicated and otherwise manipulated to create

quite complex visuals. This has been my favoured approach, specifically the spatial replication of shapes with the addition of colour and feedback effects.

Using Shape Primitives

Primitive shapes such as lines, cubes, torii, circles and so on are available in many visual packages. These simple objects can be moved, distorted and coloured over time, all to create evolving narratives of the shapes themselves. To me this is a key essence of visual music and the way such shapes can be made to behave in an almost musical manner. When coupled with music very strong cohesion between visual object and sound can be achieved. Aside from animating single primitive shapes, a useful technique is to duplicate them to create more intricate forms. TouchDesigner uses a process called 'instancing' which can also be known as duplication, replication and similar terms in other platforms. Using this specific type of instancing, or its equivalent, the visuals are synthesised on the graphics card, making the process less reliant on the computer CPU and therefore rendering more quickly and efficiently, especially if the computer has a powerful graphics card. Figure 5.4 shows the result of replicating a single hexagon (created from a torus, limited to six sides) twice, four, eight and sixteen times. Notice how the replications interact to create images like 3D boxes. This is more evident in the moving image version of these animations and this example is inspired by Whitney's *Matrix 3* (1972) film, where various repeated and animated shapes create unfolding interweaving patterns.

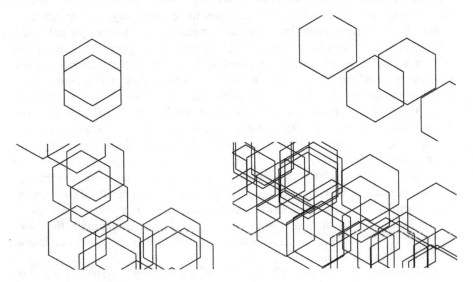

Figure 5.4 Instancing Repeated Geometric Forms. Top Row: 2 and 4 Duplications. Second Row: 8 and 16 Duplications

Video Feedback

Video feedback is a powerful effect for enhancing visuals. It can be compared to reverb and delay in sound, where repetitions of a sound can be heard some time after it was first heard. With video, images can continue to be displayed in subsequent frames after they first appeared. If an object in the frame moves, feedback creates a blend of the previous image/s with the new image on top. If this continues indefinitely, with very high feedback settings, the resultant image becomes very dense. If, however, the opacity of the feedback image is reduced, the residual image gradually fades, creating a diffuse texture and sometimes a pattern, depending on the original source image. It is common for the feedback image to be blurred, rotated, coloured or otherwise treated to create more effective results. An example of rotated feedback can be viewed on my Vimeo channel (Payling 2015a) and in Figure 5.5. In After Effects, feedback can be created using the CC Wide Time and Delay effects. Touchdesigner has a dedicated Feedback TOP.

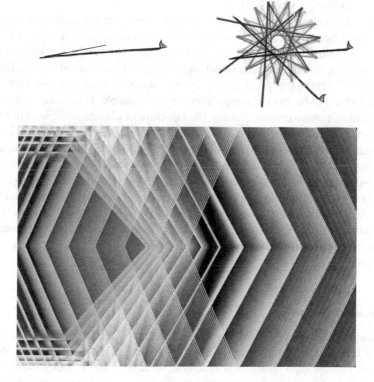

Figure 5.5 Feedback. Top Left: Original Image. Top Right: The Same Image with Rotated Feedback Trails. Bottom: A More Complex Image Created from Repeated Primitive Shapes and Feedback

Animation Techniques in Circadian Echoes

This section will discuss a case study of the development stages involved in creating the composition *Circadian Echoes* (Payling 2015b). *Circadian Echoes* is an example where the use of primitive shapes, duplication and feedback created evolving figures and visual narrative. *Circadian Echoes* was originally inspired by Norman McLaren's film *Pas de Deux* (McLaren 1968). In this film, two silhouetted dancers perform a pas de deux and their movements are enhanced by visual effects. Of particular interest to the development of *Circadian Echoes* was the dancers' movements between points across the stage, the flowing motion of their arms and McLaren's video post-production work. The aim was therefore to translate some of these attributes into an abstract composition.

The compositional process for creating the foreground imagery of *Circadian Echoes* involved developing a method of animating gestural motions of a CGI line. This was intended to create a mimetic interpretation of the dancers' arm movements. The development of this animation began with a multiple line spline curve that would randomly relocate its points in a flowing motion when a computer mouse was clicked. This can be seen in the video *Circadian Echoes: Development Stage 1 – Motion* (Payling 2016a).

The next development stage (Payling 2016b) involved the addition of feedback, which was rotated to create optical effects similar to those seen in McLaren's work. These basic animations were the foundation of the visuals of this composition. To further enhance the imagery additional processing was undertaken and multiple renders with different colours and other properties were blended together. This created circular forms and further surrogacy from the recognisable dance-like gestures.

In terms of compositional strategy, the structure of *Circadian Echoes* is a mostly consistent textural piece with a brief passage where the music reduces intensity. The intention was to create passages of tension and release, which can be used as a structural technique in music compositions and visual montages (Evans 2005). This was achieved by development of the abstract forms and tension up to 2'55", at which point they resolve into a balanced, mirrored symmetrical pattern coincident with a diminuendo in the soundtrack. This is the period of release before the tension again begins to build. A mirroring effect is used here, and in other places in the composition, and was created by a standard video mirroring technique described in the symmetry section in this chapter.

Animation Keyframes

Automation in music DAWs involves adjusting values, such as fader levels, over time, and storing these for later edit, recall and automated mixdown. In NLVEs the equivalent technique is keyframing. Keyframing means setting values for a specific parameter at one time point (frame) and different values at another frame being transitioned to. Keyframing like this is used extensively to develop intricacy in a composition. When animating with keyframes there are several ways the change from one state/frame to another can occur. A straight-line linear ramp is one possibility. This type of change, when linked to motion, can appear mechanical and create a somewhat un-natural type of movement of an object. Real-world motion often occurs in a smoother form with a gradual increase and decrease in velocity. To recreate

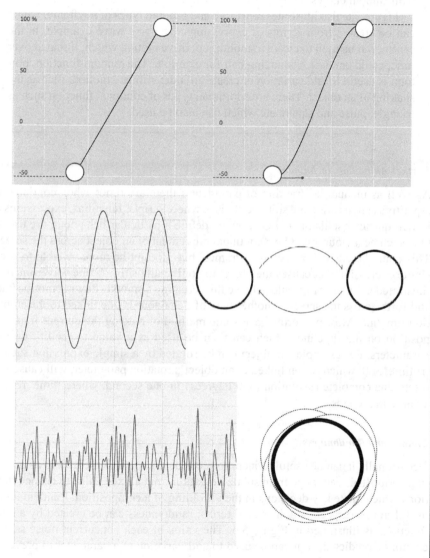

Figure 5.6 Top Left: Linear Keyframes. Top Right: Spline Curve Keyframes. Mid-dle: Sine Wave for Repetitive Motion. Bottom: Noise (Random) for Inde-terminate Behaviour

this with keyframes it is often preferable to use spline curves, which have a more flowing and natural motion. In some apps spline curves are activated using 'ease', and in After Effects 'easy ease', -type keyframes. In Touchdesigner a similar smoothing effect can be created by using a 'filter' CHOP on changing data values. More natural audio crossfades can also be achieved with smooth curves.

Figure 5.6 also illustrates repetitive and random types of keyframe, which can be created from scripts or expressions. The sine wave example, in the centre, can be used to create a smooth repetitive motion which, if scaled over time, could emulate a bouncing ball for example. The random function, bottom, is useful to add variation or create an indeterminate motion, such as the shaking of an object. There are additional types of continual function such as triangle, pulse and square etc which can also be used.

Further Animation Techniques

As well as manual keyframing of parameters there are times when continual or repetitive actions are more suitable. This is where scripts, functions, expressions or low frequency oscillators (LFOs), which define a particular behaviour, are useful. One such behaviour might be a continual movement of an object across the screen. This could be created by using keyframes, but it might be more suitable to use a single expression that causes the object to continually shift. A sine wave function, illustrated in Figure 5.6, could achieve this, creating a smooth-flowing motion back and forth across the screen. Another type of behaviour is one linked to the animation timeline. Most platforms have some method of reading the current frame, or position on the timeline, which can then be used as a value to modify desired parameters. For example, an object can be rotated by a simple expression such as 'r=time*360' which, when linked to an object's rotation parameter, will cause it to rotate one complete revolution (360 degrees) in one second, where 'time' returns a value in seconds.

Chance and Randomness

Occasionally it can be useful to incorporate indeterminate elements into an animation script. This can range from subtle deviation from a fixed value to chance functions which make key decisions in the unfolding of a composition. Randomness is useful in this respect, and, in digital terms, randomness can be created by a noise function, as illustrated in Figure 5.6. The value of each concurrent noise sample cannot be predicted, so it can be used to add variation to a parameter to prevent it becoming too static.

A complete discussion on scripting techniques is outside the scope of this text, but there are many online tutorials where such expressions can be copied and adapted. Many of these use JavaScript, Python or similar command line languages for linking and adjusting values relative to the timeline or an independent clock or counter.

Algorithmic Techniques

The techniques discussed above involve direct action and intervention by the composer. All decisions and processes are undertaken by the human artist. It is also possible to utilise computer-based algorithmic processes in composition and production, often only requiring human input at the preliminary stages. Even if they are not used to generate an entire composition, algorithmic techniques can be useful in generating ideas and unexpected source materials for later development by the composer. By taking charge and adapting the results generated from algorithmic techniques, the composer can later apply the vast array of post-processing effects available to them and transform the results in their NLVE based on their own intuitive insights in an algorithmically *assisted* manner (Battey 2021).

In some ways, using algorithmic techniques can be like working with another composer in a collaborative manner. The results produced by an algorithmic process can be unexpected and offer more potential than those achieved by manual production techniques alone. They can challenge the composer's preconceptions and take things in new and unexpected directions, as would be the case when working with others. Sometimes algorithmic techniques are used without the composer being aware, as certain effects will use computer calculated processes. Figure 5.7 visualises the use of algorithms alongside artist rendition.

Algorithms in Media Production

The left of this diagram, artist rendition, indicates where the composer is in control of all processes. A typical approach would be adding media materials to a software editor, arranging them on a timeline and applying any necessary effects. Next on the continuum is where certain automated processes are used to manipulate parameters. This could be the use of a random LFO to adjust a filter

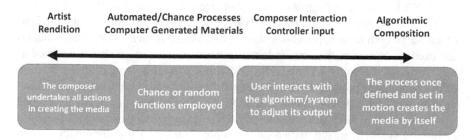

Figure 5.7 The Artist and the Algorithm. Employing Algorithmic Processes in Media Production

cutoff for example. The effect parameters are chosen and the output rendered to create various results. In this scenario the process, or algorithm, is calculating one small part of the media production. If even more control is relinquished to the *process*, *it* will determine a larger portion of the output media. Consider a modular sound synthesis rack. It can generate sound purely from the initialisation of parameters and connections between various modules. Its output sound can continue indefinitely with potentially infinite variety. It is possible, however, for the musician to adjust parameters on the fly, or change the patching arrangement, thereby producing different sound output. Finally, a totally algorithmic process is one where values are initialised, the algorithm is set in motion and the results are rendered without any further human intervention. The artist has some influence on the process during initialisation, but the computer generates all the materials based on the algorithm. As always, a combination of these techniques can be useful and productive.

Hue Music Algorithmic Composition

As an example of an algorithmic approach, *Hue Music* is introduced here. *Hue Music* is a program created in Max which interprets colours as sound timbres. The general concept was to map sound timbres to different colours, rather than colour to pitch which is more commonly used. The MAX patch in effect sonifies the colour content of a chosen image and occupies the third position of the continuum in Figure 5.7. Some interaction is possible, but it is largely the process which determines the output. A more technical discussion of this algorithm was documented in a poster presented at ICAD Montreal (Payling et al., 2007).

Image Analysis

Before converting the colours in an image to sound it is necessary to quantify this in some way. One method of achieving this is to scan through the image's pixels row by row and generate a sound that represents the colour values of each pixel. Several programs, such as Metasynth's Image Synth editor (Spiegel, 2006) and the vOICe (Meijer, 1992), use this technique. Metasynth is a creative sound design tool whereas the vOICe is used as an augmented reality device which, after a brief familiarisation period, allows sight impaired users to 'view' the environment through the medium of sound. In both these applications pixel lightness and vertical position information are used to determine the amplitude and pitch of one or more sine wave oscillators. Alexander Zolotov's Virtual ANS (Zolotov, 2014) is a similar implementation of this technique allowing advanced manipulation and editing functionality. The audible results of this type of sound generation range from randomly changing pitched sine tones to more fluid glissandi and drone-like textures. All these sounds possess a distinctive synthetic quality with a similar sonic signature. By starting with a series of more natural timbres, rather than sine wave oscillators, it could be possible to produce less synthetic sounding results, and this was the goal in creating the *Hue Music* algorithm. As an alternative to the left-right image scanning process described above, *Hue Music* also takes a more holistic approach by analysing the colour content of the currently displayed part of the image. This information is then used to determine the loudness of different timbres in the output sound.

Colour Analysis and Sound Mapping

At the heart of the colour analysis part of *Hue Music* is a process that breaks the image into an 8 x 8 grid pattern, identifies the hue value of each part of the grid and quantises it to the closest colour from a palette of black, red, yellow, green, cyan, magenta and white. Eight hues can therefore be identified with this process, and these can be mapped to different timbres. Even though eight colours are available, only seven timbres are used. In this application white is equated to silence. Using an artistic analogy, the white areas of an artist's canvas contain no colour, so they make no sound. Conversely it could be argued that white space equates to white noise, an analogy used in *Arnulf Rainer* by Peter Kubelka (1960) who also aligned silence with black, or darkness. Ultimately there is no universal one-to-one mapping and different choices can be more preferential for different images. The use of seven sounds still provides enough textures for timbral layering with more than this potentially being overwhelming or confusing. In fact, four layers at any given moment can be sufficient (Landy 2007, 69). Furthermore, analysis of more colours would diminish the elegance of the use of the two well-known RGB and CMYK colour systems in the chosen palette.

As an example of how the analysis and mapping operates, consider a simple image which is exactly half blue and half yellow. *Hue Music* automatically splits it into an 8 x 8 grid, giving a total of 64 squares. The colours in these grid 'squares' are counted and, in this case, there would be 32 blue squares and 32 yellow squares. These numbers are then fed into the sound generation routine that produces a sound that is a combination of the blue and yellow timbres at equal amplitude. Fifty per cent of the image is blue, therefore 50 per cent of the sound is 'blue'. A typical photograph would contain a wider range of colours, producing a more complex combination of timbres, but the maximum number of available squares is always 64. A photograph would also commonly contain colours outside the available primary palette colours. In this case another process of the *Hue Music* algorithm quantises all colours to the one closest in the available palette.

Consider therefore a photograph with multiple colours being loaded into the program. As above, the 64 grid squares are quantised and counted to determine the quantity of each hue value. In this particular image a greater selection of the palette is represented so that the process evaluates there are 37 black squares, 9 green squares, 1 blue square, 17 cyan squares and the remaining colours are not present. Note again that this sums to 64. These colour counts are then used to adjust the timbral mixer described below, and in this case the black timbre would be dominant as it has the largest count, followed by cyan, green then blue.

Audio Timbre Mixer

The music is produced by an automated mixer controlled by the colour content of the image. Each channel of the mixer is assigned to a colour and each colour has an associated timbre. The timbres are pre-recorded audio files loaded into the program. Each of the timbres were recorded by the composer using a variety of found sounds and synthesisers and saved as files around 30 seconds in duration, each with some variation. One file might be 30 seconds long, whereas the next might be 32 seconds. The sound files loop repeatedly and are increased or attenuated in the mix by the automated mixer. Different length sound files ensures some variation in the overall soundscape, as sections which overlap change gradually over time due to having different loop points. Consider Eno's *Music for Airports'* track 2/1 (1978). This piece is created from tape loops of differing lengths, each containing a different sung note.

> One of the notes repeats every 23 ½ seconds. It is in fact a long [tape] loop running around a series of tubular aluminum chairs in Conny Plank's studio. The next lowest loop repeats every 25 7/8 seconds or something like that. The third one every 29 15/16 seconds or something. What I mean is they all repeat in cycles that are called incommensurable-they are not likely to come back into sync again.
>
> (Eno 1996)

This creates an evolving harmonic and timbral arrangement with a lot of variety from a relatively simple range of source sounds.

The timbres discussed above were conceived and recorded by the composer attempting to create sounds that resembled the chosen colour. This was a very subjective process and was based partly on the literature reviewed in Chapter 2 (Gerstner 1981; Jones 1973; Kandinsky 1914), and partly on the composer's own intuitive alignments. What works for one composer could be completely at odds to another's. The priority in this case was that the resulting music contained the desired qualities in relation to the picture being converted. In this case black and blue were considered to align with deep, bass-orientated sounds and yellow and green were brighter sharper sounds. The synthesiser and sampler parameters were edited until the timbre matched the colour being associated.

Timbre Amplitude Adjustment

The values calculated from the colour analysis are now used to adjust the timbral mixer. The maximum number available is 64. A value of 64 for any hue would mean that the associated timbre was played at maximum amplitude, in which case only one sound is heard, as this image would be monochromatic. Referring also to the analogy of white space being equal to no sound, a completely white image would create silence.

The process can be imagined as a seven-channel stereo audio mixer. Each of the stereo samples is continuously looped and their amplitudes adjusted by the calculated colour content in the previous example, with 37 black squares, the timbre associated with black would be the loudest. The green, blue and cyan timbres are also audible, but at a reduced amplitude.

Scanning and Focus

By converting to timbres only the colour quantities in the picture, no sense of the spatial composition of the picture is conveyed. This is resolved by the ability to zoom into a particular area in the image. This is another implementation of the Ken Burns effect, discussed earlier in this chapter. In this case, by scanning across various parts of the picture, the timbre will change as different colours are emphasised and give auditory clues as to the visual composition of the picture. Without this zoom and scan feature, the timbral soundscape would be quite static. Some variety would be produced because of the varying lengths of the looped sound files, the dynamic quality of the recordings and the way different timbres interact. There would, however, be little in the way of variation over longer periods and no interaction.

Hue Music Discussion

Hue Music will generate similar music for all images containing similar colour content. This is because only the colour content is converted audio timbres, the algorithm does not take into account figures and forms in a picture. This means the included timbres are suitable for some scenes but not others. To remedy this issue it is up to the user to record their own sound files which will be more appropriate for their chosen purpose. For those interested in further modification of the Max patch there are several things to consider modifying. To sonify the visual compositional structure of the image and produce a musical structure based on this, the scanning process could trace a fixed path across the entire picture, so every section of the image is sonified in a predictable manner. Different images would create different musical arrangements even if the timbres remained the same as the colours would be in different spatial positions in the image. This would, however, create a repetitive loop. A more thorough analysis of the image could also be used to control an alternative timbre mixer. If important figures were identified in the picture, they could be represented by a strong timbral focus or identifiable musical composition themselves. This would require the isolation of figures from the background. If this were the case, background colours and textures could be assigned reduced amplitude in the mix or made more ambient in nature. This would follow a foreground and background approach to mixing where important sounds are emphasised against those that are fulfilling a less important role. Additional manipulation of the pitch of the timbre files based on the colour saturation, such as that described by Giannakis (2001), could also potentially enhance the variety of the generated music. Giannakis determined the most approximate mapping between colour values and sound was to associate brightness with pitch and saturation with loudness. This means a heavily saturated brightly coloured square would be louder and higher in pitch than a dull grey square. Another variation would be to use a video as the visual source instead of a still image. A video naturally contains its own time base and changing colour content which will produce a changing composition over time. Similarly, instead of an existing image or video, the interface could be modified to allow a user to draw the image themselves, making this software a performance and composition tool.

Hue Music Summary

Hue Music is a means of using the colour content of an image to create music through automated timbral mixing. As previously discussed, and with reference to the variations in associations experienced by synaesthetes (Motluk 1996), there is no universal mapping between one timbre and one colour which is appropriate for all people. It is the context of the colours in the picture and how they map to the current timbral palette that determines whether the result is successful.

Hue Music is just one example of an algorithmic technique for generating sound from an image. There are many more such approaches and algorithms which can be created by the composer themselves or third-party applications. One recent development in algorithmic processes is the use of artificial intelligence to create sound and image from descriptive text. Music applications such as Open Jukebox for music can produce convincing musical results purely from code and are being continually adapted and improved.

Sound Processing and Synthesis Techniques

So far this chapter has mostly examined visual development techniques. To complement this I will review some of the sound processing techniques that I have found useful in music composition.

Tape Techniques

First are traditional 'tape' techniques involving playback rate adjustment, reversing and crossfading. These simple techniques are often overlooked in favour of more complex plugin processes, but they can be very effective. There are various ways sound files can be time-stretched and many platforms use different algorithms designed for different categories of sound. Ableton Live allows the selection of different 'warp' methods involving the elastique algorithm, granular type processing or 'tape' style re-pitching. Vastly different results can be achieved through the different methods and the settings therein. As well as applying fixed stretch values, consider automating them. Sometimes it is necessary to perform this in real-time while the results are recorded. Imagine continually pulling and reversing a length of pre-recorded magnetic tape over a tape head and the results you might achieve by doing this. How can this be recreated digitally? The application 'Paul Stretch' (Sonosaurus n.d.) is another excellent tool for time-stretching audio files. It is best used for creating sound textures and atmospheres as it stretches audio beyond all reasonable limits. A related technique is to reverse a sample. This can reduce the percussive elements in transient sounds, for example, and coupled with time-stretching can create other unexpected results.

FFT Based Processing

The Fast Fourier Transform (FFT) is a mathematical process which can determine the frequency and phase content of an audio file. It is commonly used in spectral analysis tools but can also be found in filtering effects, especially where precise or brick-wall filtering is required. The Michael Norris Soundmagic Spectral Plugins (Norris 2014) are a set of 23 free plugins (Mac only, audio units) that use FFT based spectral processing to produce an extensive range of sounds suitable for electronic composition. The plugins have *many* parameters, some of which are duplicated across all the plugins, such as FFT size and Lo and Hi bin cutoff frequencies. Others are specific to the plugin and require some experimentation. The plugins can be used for static fixed parameter type processing, but they can also be automated to create dynamically shifting results. The plugin Spectral Freeze, for example, captures a block of the spectrum and can repeat it indefinitely to create sound textures. This plugin can be adjusted and automated to produce sounds ranging from short reverbs to infinite drones. Spectral Drone Maker is good for use on noise-based sources as it applies a pitched comb filter bank thereby creating drones in various musical scales. This plugin is useful when manipulated in real-time. By creating a gestural interface allowing real-time interaction with the plugin, coupled with recording capability, this interface and the Drone Maker effect were used extensively in the soundtrack for *Diffraction* (Payling 2012).

Granular Synthesis

Whereas tape techniques can be explained by sound wave theory, granular synthesis techniques are more analogous to quantum-based particle physics. Granular synthesis works with short fragments of sound that are reproduced at varying rates to create clouds of sound particles. The plugin

> Granulator II is a Max4Live synthesizer based on the principle of quasi-synchronous granular synthesis. It creates a constant stream of short cross-fading sections of the source sample, and the pitch, position and volume of each grain can be modulated in many ways to create a great variety of interesting sounds.
>
> (Henke, 2016)

Granulator II is a free download (Henke 2016) but it only runs in Ableton Live as a Max for Live instrument. It has an intuitive interface and can be used to create textural sounds but will also allow the creation of glitchy-type effects depending on the playback parameters and source sound. Containing similar features, but with a more extensive range of controls, is EmissionControl2. This is a 'studio grade' granular processing tool and is free and cross-platform, available for OSX, Windows and Linux (Roads, Kilgore, and DuPlessis 2022).

Convolution

Convolution of two audio files can be achieved by multiplying together their FFT spectra. Typically, when using convolution on sound files, one of the spectra would be from a real reverb tail which can be created by stimulating a space, such as a room, with an impulse response. Convolution is therefore commonly used in reverb plugins where natural sounding reverb effects are desired. It is, however, possible to use any type of sound file as the 'impulse response' to create complex timbral effects from originally mundane type recordings. Try loading a short sound file into a convolution reverb effect such as Ableton and Adobe Audition's Convolution Reverb or Logic's Space Designer and playing a variety of sounds through the effect.

Sound Synthesis

As discussed above, sound synthesis is a super flexible technique for designing sound that will fit very closely with moving images. It permits precise programming and automation so that sound actions can be made to mimic those occurring in the visuals. It requires a very different approach to composition than that when utilising techniques derived from musique concrète. There are several ways into using synthesis with possibly the easiest route being through software techniques. Most modern DAWs have a large selection of synthesiser plugins and there are many third-party plugins which will run in a host DAW. For those with the desire for increased customisation and control consider using a sound coding platform such as Supercollider or Max. Use of these types of tools will go a long way to developing an understanding of the fundamentals of sound synthesis which can then be translated into other devices, increasing synth programming skills and enabling rapid tweaking of presets. In hardware, Eurorack-based modular systems are very popular and permit gradual expansion of a system when funds and space permit. This gradual expansion gives the user time to learn individual modules as they are acquired. Several virtual modular systems are also available such as VCV Rack (Belt 2022) which is available as a free version with paid upgrades. Of course, there is a massive range of hardware synths, drum machines, sequencers and other devices which give a more tactile and ergonomic experience than computer-based systems, the main limits being space and finances.

One software synthesis tool that keeps returning to use in my productions is Reaktor. It can be used for anything from experimental forms of sound synthesis through to commercial beat-based EDM styles. Reaktor is a programming environment for building synthesisers and audio effects. As well as its synthesis capabilities it also has sampling and granular processing functionality for working with sound files. It takes a low-level modular-based programming approach with signal processing and logic blocks connected by virtual cables. When compared to Max it is less focused towards control and logic but it makes up for this with a high-quality synthesis engine and intuitive interface which can make synth and effects building more straightforward. Newer versions also contain a range of virtual modular synthesis devices which can be linked with virtual patch-chords to create a digital modular synthesis rack. Reaktor has such an extensive range of prebuilt library devices it could be the only software plugin one would need, so it is not necessary to use its device-building capabilities unless you have a specific function in mind which is not available.

Summary

This chapter has introduced techniques for creating audio-visual materials to aid construction visual music compositions or elements to be used in live performance. These are intended to be suggestions and starting points from which the reader can expand upon or use as inspiration for their own methods. They are not intended to be hard rules which should be copied, but a launchpad for the individual to develop their own ideas and artistic style. There are many more approaches and alternatives to the techniques discussed above. It will be of great interest to see how these have inspired other artists and I am curious to see the results of their creativity.

Note

1 The discussion in this chapter has more emphasis on composition but many of the techniques can be used for performance purposes.

References

Battey, Bret. 2021. *Fluid Audiovisual Counterpoint in Estuaries 3 - Bret Battey (EUA)*. www.youtube.com/watch?v=dN5I.9-DMl5w.
Belt, Andrew. 2022. 'VCV Rack—The Eurorack Simulator for Windows/Mac/Linux'. https://vcvrack.com/.
Boucher, Miriam and Jean Piché. 2020. 'Sound/Image Relations in Videomusic: A Typological Proposition'. In *Sound and Image: Aesthetics and Practices*, edited by Andrew Knight-Hill, p. 414. London: Focal Press. www.taylorfrancis.com/chapters/edit/10.4324/97804292951 02-2/sound-image-relations-videomusic-myriam-boucher-jean-pich%C3%A9.
Brian Eno. 1978. *Ambient 1: Music for Airports*. Polydor. www.discogs.com/release/125776-Brian-Eno-Ambient-1-Music-For-Airports.
Burns, Ken. 2012. *The Civil War: The Cause*. www.youtube.com/watch?v=FN2huQB-DmE.
Chion, Michel. 1983. *Guide Des Objets Sonores. Pierre Schaeffer et la Recherche Musicale*. Translated by John Dack and Christine North. Translated from French. Paris: Institut National de l'Audiovisuel & Éditions Buchet/Chastel.
———. 1994. *Audio-Vision: Sound on Screen*. New York: Columbia University Press.
Cope, Nick and Tim Howle. 2003. *Open Circuits*. Electroacoustic Movie. https://vimeo.com/633346.
Csikszentmihalyi, Mihaly. 2013. *Creativity: The Psychology of Discovery and Invention*. Reprint edition. New York: Harper Perennial.
Eno, Brian. 1996. 'Generative Music'. *Motion Magazine*, 8 June 1996. https://inmotion magazine.com/eno1.html.
Evans, Brian. 2005. 'Foundations of a Visual Music'. *Computer Music Journal* 29 (4): 11–24.

Garro, Diego. 2012. 'From Sonic Art to Visual Music: Divergences, Convergences, Intersections'. *Organised Sound* 17 (02): 103–13. https://doi.org/10.1017/S1355771812000027.

Gerstner, Karl. 1981. *Spirit of Colour: Art of Karl Gerstner*. Cambridge, MA: MIT Press.

Giannakis, Konstantinos. 2001. 'Sound Mosaics: A Graphical User Interface for Sound Synthesis Based On Auditory-Visual Associations'. School of Computing Science: Middlesex University.

Henke, Robert. 2016. 'Granulator by Robert Henke'. Robert Henke. 11 March 2016. http://roberthenke.com/technology/granulator.html.

Hyde, Joseph. 2009. *End Transmission*. Audiovisual Composition. https://vimeo.com/3241660.

ITU. 2012. 'RECOMMENDATION ITU-R BS.1771–1—Requirements for Loudness and True-Peak Indicating Meters'. International Telecommunication Union. www.itu.int/dms_pubrec/itu-r/rec/bs/R-REC-BS.1770-3-201208-S!!PDF-E.pdf-n0tpNw&ust=1656506191137862.

Jones, Tom Douglas. 1973. *Art of Light and Colour*. New York: Van Nostrand Reinhold.

Kandinsky, Wassily. 1946. *Concerning the Spiritual in Art*. The Solomon R. Guggenheim Foundation for The Museum of Non-Objective Painting. New York City

Kanellos, Emmanouil. 2018. 'Visual Trends in Contemporary Visual Music Practice'. *Body, Space & Technology* 17 (1): 22–33. https://doi.org/10.16995/bst.294.

Kubelka, Peter. 1960. *Arnulf Rainer*. Short. www.imdb.com/title/tt0377390/.

Landy, Leigh. 2007. *Understanding the Art of Sound Organization*. Cambridge, MA: MIT Press.

Long, Ben, and Sonja Schenk. 2011. *The Digital Filmmaking Handbook*. 4th revised edition. Boston, MA: Delmar Cengage Learning.

Max Cooper. 2020. *Behind the Scenes: Making 'Circular' with Páraic McGloughlin*. www.youtube.com/watch?v=h98QtAF4uJw.

McGloughlin, Kevin and Max Cooper. 2019. *Max Cooper—Circular*. https://vimeo.com/372085570.

McLaren, Norman. 1968. *Pas de Deux*. Animation, Short, Musical. www.imdb.com/title/tt0063417/.

Meijer, Peter. 1992. 'An Experimental System for Auditory Image Representations'. IEEE Transactions on Biomedical Engineering 39 (2): 112–21. https://doi.org/10.1109/10.121642.

Motluk, Alison. 1996. 'Two Synaesthetes Talking Colour'. In *Synaesthesia: Classic and Contemporary Readings*, edited by John E. Harrison and Simon Baron-Cohen, 269–77. Oxford: Wiley-Blackwell.

Norris, Michael. 2014. 'Soundmagic Spectral'. Michael Norris: Composer, Software Developer, Music Theorist. www.michaelnorris.info/software/soundmagic-spectral.html.

Paterson, Ian. 2004. *A Dictionary of Colour: A Lexicon of the Language of Colour*. London: Thorogood.

Payling, Dave, Stella Mills and Tim Howle. 2007. 'Hue Music–Creating Timbral Soundscapes from Coloured Pictures'. In *The International Conference on Auditory Display*, Quebec, Canada, 91–97. Montreal.

———. 2012. *Diffraction*. H.264 Electronic Visual Music. https://vimeo.com/40227256.

———. 2015a. *Quartz Composer Feedback Experiment (No Sound)*. Electronic Visual Music. https://vimeo.com/122754371.

———. 2015b. *Circadian Echoes*. H.264 Electronic Visual Music. https://vimeo.com/127819711.

———. 2016a. *Circadian Echoes—Development Stage 1—Motion*. Electronic Visual Music. https://vimeo.com/173621178.

————. 2016b. *Circadian Echoes—Development Stage 2—Creating Delay with Feedback.* Electronic Visual Music. https://vimeo.com/173622726.

Piché, Jean. 2011. *Océanes.* Video Music. https://vimeo.com/25933560.

Roads, Curtis. 1996. *The Computer Music Tutorial.* Cambridge, MA: MIT Press.

Roads, Curtis, Jack Kilgore and Rodney DuPlessis. 2022. 'EmissionControl2: Designing a Real-Time Sound File Granulator'. In *Proceedings of the ICMC*, 49–54. Limerick: International Computer Music Association, Inc.

Sharits, Paul. 1966. *Ray Gun Virus.* Short.

————. 1968. *T,O,U,C,H,I,N,G.* Short.

Smalley, Dennis. 1986. 'Spectro-Morphology and Structuring Processes'. In *The Language of Electroacoustic Music*, edited by Simon Emmerson, 61–93. London: Palgrave Macmillan.

Sonosaurus. n.d. 'PaulXStretch'. Accessed 2 August 2022. https://sonosaurus.com/paulxstretch/.

Soundfield. 2023. 'SoundField | Microphones and Processors with Unique Surround Sound Capabilities'. www.soundfield.com.

Spiegel, Edward. 2006. 'Metasynth 4.0. User Guide and Reference'. U&I Software, LLC. https://uisoftware.com/metasynth-manual/.

Weyl, Herman. 1952. *Symmetry.* Princeton, NJ: Princeton University Press.

Whitney, John. 1972. *Matrix III.* Animated Short. www.imdb.com/title/tt2600366/.

Wishart, Trevor. 1996. *On Sonic Art.* Edited by Simon Emmerson. 2nd revised edition. Amsterdam: Routledge.

YouLean. 2023. 'Youlean Loudness Meter—Free VST, AU and AAX Plugin'. Youlean. 2023. https://youlean.co/youlean-loudness-meter/.

Zolotov, Alexander. 2014. 'WarmPlace.Ru. Virtual ANS Spectral Synthesizer'. WarmPlace. Ru. http://www.warmplace.ru/soft/ans/

Index

Printed in the United States
by Baker & Taylor Publisher Services